G000125965

MATHEMATICS RESEARCH DEVELOPMENTS

DETERMINISTIC AND RANDOM EVOLUTION

MATHEMATICS RESEARCH DEVELOPMENTS

Additional books in this series can be found on Nova's website
under the Series tab.

Additional e-books in this series can be found on Nova's website
under the e-book tab.

MATHEMATICS RESEARCH DEVELOPMENTS

DETERMINISTIC AND RANDOM EVOLUTION

JENS LORENZ
EDITOR

New York

NOTICE TO THE READER

Library of Congress Cataloging-in-Publication Data

ISBN: 978-1-62618-014-7

Published by Nova Science Publishers, Inc. † New York

Contents

Mathematics is sharpening of common sense.

Preface

The first notes for this text were written during the summers of 2008–2010 when I taught a short course on mathematical modeling at the University of New Mexico. The audience consisted mostly of undergraduate mathematics students, and an aim of the course was to interest them in math at the graduate level.

The students had some basic knowledge of ordinary differential equations and numerics. I tried to build on this foundation, but instead of increasing the technical skills of the studen I tried to lead them to more fundamental questions. What can one model with differential equations? What is determinism? If the universe evolves deterministically, what about free will and responsibility? Does it help if there are elements of randomness in the laws of evolution?

Of course, these are deep questions, and in this text we can only scratch the surface in our discussion. Nevertheless, mathematics — even at a rather elementary level — may help to clarify what is at stake. Throughout, I try to put the discussion into historical context. For example, the text contains a rather detailed description of the derivation of Kepler's laws of planetary motion using ordinary differential equations. After all, Newton's great success in deriving these laws were an important starting point of the scientific revolution and a deterministic world view. It made classical mechanics a model for all sciences.

Even if a deterministic description of an evolution is possible, there are often practical limitations of predictability because of the exponential growth in time of any uncertainty in the initial condition. Iteration with the logistic map gives an example. However, even if the accurate determination of future states is impractical, the *average behavior* of a system may still be very robustly determined. The logistic map again serves as an example. Do we have a similar situation for weather and climate? We cannot predict the weather two weeks in advance, but it may still be possible to determine the average weather 30 years from now.

There are similarities to random evolution. When throwing a fair coin, we cannot predict the outcome of the n–th throw, but we can be rather certain to have between 450 and 550 heads in a thousand throws. If you do not want to compute the exact probability for this claim and also do not want to throw a coin many thousand times, you can test the claim using a Matlab code and Matlab's random number generator. Some simple Matlab codes are provided in the text. They may encourage the readers to get their own experience with models for deterministic or random evolution.

The mathematical level of the text corresponds to the ability and experience of undergraduate mathematics students making the critical transition to graduate work. The text is accessible if you had an undergraduate course on ordinary differential equations and numerical methods. For some parts, it is good to be familiar with elementary concepts of probability and statistics, though the concepts will be reviewed in the text.

The course was part of an MCTP program, *Mentoring through Critical Transition Points*, supported by the NSF. It is a pleasant task to acknowledge the support by the NSF and to thank the PIs of the grant, Prof. Cristina Pereyra and Prof. Monika Nitsche, for their tireless work on all the details of the MCTP program. I also like to thank Katherine and Randy Ott, David Phillips, and Olumuyiwa Oluwasami for proof reading the text, for making figures, and for managing Latex.

This material is based upon work supported by the National Science Foundation under Grant No. 0739417.

Jens Lorenz

Department of Mathematics and Statistics,
University of New Mexico,
Albuquerque, New Mexico, US
E-mail address: lorenz@math.unm.edu

Chapter 1

Introduction

Summary: First we briefly introduce three classes of models, two of them are deterministic, namely initial value problems for ordinary differential and difference equations. The third class of models is random evolution in discrete time.

We then comment on three subjects that will be developed further in the text:

1. Newton's derivation of Kepler's laws and the resulting deterministic world view.

2. For sensitive deterministic systems, it is practically impossible to predict individual solutions, but *averages* may still be robustly determined. Is the weather/climate system an example?

3. Kinetic theory as an example relating micro and macro models.

Evolution is the change over time. In this text we introduce the reader to three classes of models that are used to describe evolutionary processes:

a) initial value problems for ordinary differential equations:

$$\frac{du}{dt} = f(u), \quad u(0) = u_0 \; . \tag{1.1}$$

Here $u(t)$ is a vector that specifies the state of a system at time t. At time $t = 0$ the state vector is given by the initial condition $u(0) = u_0$, and the differential equation $\frac{du}{dt} = f(u)$ then determines how the state $u(t)$ evolves in time.

The function $f(u)$ in (1.1) often depends on a parameter λ, i.e., $f(u) = f(u, \lambda)$. If this is the case, bifurcations and hysteresis phenomena may occur. Using simple examples, we will introduce these subjects in Chapter 8.

b) difference equations:

$$u_{n+1} = \Phi(u_n), \quad n = 0, 1, 2, \ldots \tag{1.2}$$

The vector u_n specifies the state of the system at time $t = n$, and Φ is a map of the state space into itself, which is assumed to be known. Given an initial state u_0, the equation (1.2) determines the sequence of states

$$u_0, u_1, u_2, \ldots$$

One calls $\{u_n\}_{n=0,1,2,\ldots}$ the orbit with initial value, or seed, u_0.

Whereas time is a continuous variable in (1.1), it changes discretely in (1.2). Time discretization occurs, for example, if one applies numerical codes to solve (1.1).

In general, replacing continuous time by discrete time, changes the dynamics not only quantitatively but also qualitatively. We will discuss this change using the logistic differential equation and the delayed logistic map as examples in Chapter 8.

c) random evolution: We will consider processes of the form

$$u_n \to u_{n+1} \quad \text{with probability} \quad p = p(u_n, u_{n+1}), \qquad n = 0, 1, 2 \ldots \tag{1.3}$$

The gambler's ruin problem serves as an example for the process (1.3); see Chapters 11 and 12. Random evolution with continuous time will also be discussed. In Chapter 13 we illustrate this by a model for a stochastic growth process.

We will apply elementary concepts from probability theory to analyze random evolutions and then compare the analytical results with numerical simulations. Matlab's random number generator is a great tool for running such simulations.

Throughout the text, simple Matlab codes are provided. They were not written to emphasize speed of execution or elegance of coding. The only aim is to make them easy to read. This may encourage readers to modify the codes and develop their own experience in mathematical modeling of evolutionary processes.

An aim of the text is also to discuss the more fundamental, or philosophical, aspects of mathematical models for evolution. A key figure of the scientific revolution of the 17th century was Johannes Kepler (1571–1630) who formulated his three laws for planetary motion in 1609 (first and second law) and 1619 (third law). Almost a century later, Isaac Newton (1643–1727) was able to deduce Kepler's laws by applying the inverse square law of gravitation to the two–body problem. This tremendous success of Newton not only removed the last doubts about the heliocentric system, but also advanced the scientific revolution in general. Newton's equations for the two–body problem can be formulated as an initial value problem of the form (1.1), which is a *deterministic* system (under mild assumptions on the function $f(u)$). In fact, one may say that Newton's success was a main reason for a *deterministic world view*, a search for a set of universal laws that rule the evolution of the universe. Throughout the text we will comment further on such issues. Of course, we cannot settle the age–old controversies regarding determinism, free will, causality, predictability etc. Nevertheless, the three classes of mathematical models (1.1)–(1.3) may help to clarify what is at stake.

How one models a phenomenon depends on the *scale* one is interested in. A good, but difficult, example is given by weather and climate. Weather and climate are similar phenomena, but on different time and space scales. We know the equations governing the change of the weather well and, with the help of computers, can use the equations to rather reliably predict the weather for a few days.

However, the weather system is too sensitive to make accurate predictions for a month in advance. How, then, can we predict the climate 30 years from now? This is difficult, indeed. Between day and night, the temperature at any place may change by 20 degrees or more. It may also vary by 20 degrees within a distance of less than 50 miles. This, together with the sensitivity of the weather system, makes it hard to believe that we can indeed predict a rise of the average temperature by three degrees in the next 30 years.

To illustrate that such a prediction is not completely hopeless and absurd, we will consider the logistic map, an example of a difference equation (1.2); see Chapter 6. The example shows that it may be practically impossible to predict future states accurately, but that the *average behavior* of the evolution is nevertheless very well determined and predictable. It is an open question if the same applies to the weather/climate models currently in use.

Conceptually, the situation is similar but easier for micro and macro models of fluids and gases. It is easier since we have good experiments and equations on the micro and the macro scale. Nevertheless, it is by no means trivial to connect the micro and macro models. This is the subject of kinetic theory. In Chapter 14, we give an introduction to this vast subject. If we had a similar theory relating weather and climate, our current climate predictions would have a sounder scientific base.

It seems that we face somewhat similar mathematical difficulties in micro and macro economics. The models work on different scales, but it remains a challenge to connect them mathematically with statistical arguments. How do the actions of a 100 million people determine the dollar/euro exchange rate?

In kinetic theory, the historically first result is attributed to Daniel Bernoulli (1700–1782), who tried to explain Boyle's law ($pV = const$) from a particle point of view. Bernoulli's simplifying assumptions are certainly wrong, but are still a stroke of genius and made his derivation possible. We present his arguments in Section 2.. We also describe James Clark Maxwell's (1831–1879) derivation of his velocity distribution, which is based on plausible symmetry assumptions for the probability distribution function. Maxwell's distribution, dating back to 1860, is historically very important: The first introduction of probability into physics.

A third remarkable result goes back to Sadi Carnot (1796–1832). Around 1770, James Watt (1736–1819) greatly improved the steam engine, an important starting point of the industrial revolution. Watt's engineering work also led to fundamental scientific questions: How can heat be used to produce mechanical work? Is there a most efficient way to do this? These are, of course, still important topics of engineering today. Amazingly, Sadi Carnot described an *idealized heat engine* that had *optimal efficiency*. Carnot's insights are remarkable since at his time *conservation of energy*, i.e., the first law of thermodynamics, was not yet understood.

The dynamic interpretation of Carnot's ideas later led to various formulations of the second law of thermodynamics. Some of the related mathematical issues were first addressed by Ludwig Boltzmann (1844–1906) and are still an active research topic today. In the last chapter of the book we describe a random evolution which illustrates the transition from order to chaos, or the direction of time, in an elementary way.

Chapter 2

Basic Concepts

Summary: In this chapter we will give more details about the three classes of models, initial value problems for ODEs, difference equations, and discrete–time random evolution. The important concept of a Markov chain will be introduced. We also discuss the notion of randomness.

1. Initial Value Problems for ODEs

In this section we introduce some basic concepts for first order systems of ordinary differential equations (ODEs). More advanced concepts, like stability, bifurcation, different time scales, and sensitive dependence on initial conditions will be discussed later.

Let us start with a simple example of an initial value problem for a scalar ODE:

$$u'(t) = -u(t) + 1, \quad u(0) = 3 . \tag{2.1}$$

Here $u = u(t)$ is the unknown function. We will think of the independent variable t as time. As one learns in any introductory ODE course, the general solution of the differential equation $u' = -u + 1$ is $u(t) = 1 + ce^{-t}$ where c is a free constant. Imposing the initial condition $u(0) = 3$ leads to $c = 2$, i.e., the solution of the initial value problem (2.1) is

$$u(t) = 1 + 2e^{-t} .$$

For this simple example, the following is quite obvious: The general law of evolution, expressed by the differential equation $u' = -u + 1$, together with the initial condition $u(0) = 2$, determines the value of the solution $u(t)$ exactly at any other time t. The future ($t > 0$) as well as the past ($t < 0$) values of the solution are exactly determined by the initial condition and the general law of change. In other words, the initial value problem (2.1) gives a simple example of how to encode a deterministic evolution.

In more advanced courses on ODEs one learns how to generalize these simple observations. The function $u = u(t)$ may be vector valued, $u(t) \in \mathbb{R}^N$, where $\mathbb{R}^N$ is called the state space . The general law of evolution becomes a system of ODEs, $u' = f(u)$, where $f : \mathbb{R}^N \to \mathbb{R}^N$ maps the state space into itself. [1] If $u_0 \in \mathbb{R}^N$ is any fixed state, then the initial value problem

$$u'(t) = f(u(t)), \quad u(0) = u_0 \tag{2.2}$$

is a generalization of example (2.1). If the function f obeys some technical condition [2] then the initial value problem (2.2) can be shown to have a unique solution $u(t)$ defined for all time, $-\infty < t < \infty$. As in example (2.1), the state of the system, described by $u(t)$, is precisely determined by the general law of evolution $u' = f(u)$ and the state of the system at any particular time, for example by the initial condition $u(0) = u_0$.

The clear distinction between a general law of change, like $u' = f(u)$, and the specification of the state of a system at a particular time, $u(0) = u_0$, appears to be due to Isaac Newton. Some people have called this Newton's greatest discovery. [3] One of Newton's important contributions to science is the solution of Kepler's problem. Based on observations by Tycho Brahe (1546–1601), Kepler formulated his three laws of planetary motion : (1) Every planet travels around the sun in an elliptical orbit with the sun at one focus; (2) The line joining the sun to the planet sweeps out equal areas in equal times; (3) If T is the time of revolution around the sum and a is the major semi–axis of the planets orbit, then $T^2 = const \cdot a^3$ where the constant is the same for all planets.

Newton succeeded in deriving Kepler's laws from the inverse square law of gravity and Newton's second law, *force = mass times acceleration.* Newton's success was very influential for the history of science. Classical mechanics became a model for all sciences. For example, numerous (unsuccessful) attempts were made to explain light and electro–magnetic phenomena in terms of mechanics.

Remark: The result that the initial value problem (2.2) has a unique solution if f is Lipschitz continuous can be shown by a contraction argument applied to Picard's iteration:

$$u^{n+1}(t) = u_0 + \int_0^t f(u^n(s))\, ds, \quad n = 0, 1, 2, \ldots$$

with $u^0(t) \equiv u_0$. In general, if the vector function $f(u)$ is nonlinear, one cannot obtain an explicit formula for the solution $u(t)$, however.

[1] One can also allow $f = f(u,t)$ to depend explicitly on time, i.e., the law of change may change in time.

[2] A sufficient condition is Lipschitz continuity of f, i.e., there is a constant L so that the estimate $\|f(u) - f(v)\| \leq L\|u - v\|$ holds for all $u, v \in \mathbb{R}^N$. Here $\| \cdot \|$ is any norm on the state space $\mathbb{R}^N$.

[3] This distinction has proved to be very useful in many areas of science. An exception is the cosmology of the early universe where the distinction between the initial state and the general law of evolution gets blurred.

2. Discrete–Time Dynamics

In ordinary differential equations, as considered in the previous section, the time variable t varies continuously. A phenomenon is often mathematically easier to understand if one makes t a discrete variable and considers the evolution determined by a map $\Phi : M \to M$. Here, at first, M may be any set, called the state space. The map $\Phi : M \to M$ and any fixed initial state $u_0 \in M$ then determines the sequence $\{u_n\}_{n=0,1,2,\ldots}$ by iteration:

$$u_{n+1} = \Phi(u_n), \quad n = 0, 1, 2, \ldots \tag{2.3}$$

If we think of the nonnegative integer n as time, then the map Φ determines how the initial state u_0 evolves in forward time. Here we have a deterministic evolution into the future, but not into the past.

A simple example is given by the so–called logistic map where

$$M = [0, 1], \quad \Phi(u) = 4u(1 - u) \quad \text{for} \quad 0 \le u \le 1 \ .$$

The study of dynamical systems (2.3) becomes mathematically interesting if one puts some structure on the state space M and the map Φ. For example, *ergodic theory* deals with the case where M is a measure space and Φ a measurable map. In this text we restrict ourselves to the case where $M = \mathbb{R}$ is the real line, or M is some subinterval of the real line, or $M = \mathbb{R} \bmod 1 = S^1$ is the unit circle.

Our main reason for studying the evolution determined by maps is simply that the mathematics is often easier for a map than for an ODE.

3. Continuous vs. Discrete Time

To model a phenomenon, should one use a continuous–time ($u' = f(u)$) or a discrete–time ($u_{n+1} = \Phi(u_n)$) description? Traditionally, continuous–time descriptions by ordinary or partial differential equations have been preferred, in particular in physics. The resulting models are more easily accessible to mathematical analysis. On the other hand, discrete–time models are directly suitable for computer simulations. They may also be more natural in biological applications, for example for modeling growth processes from one generation to another.

Many differential equations cannot be solved analytically. One then uses numerical algorithms to approximate the solution, which — in some sense — means that one replaces continuous time by discrete time. Numerical analysis deals with the error between the continuous–time analytical model and the result of the numerical approximation.

In Chapter 8 we consider the continuous–time growth model of logistic growth and also a corresponding discrete–time model. We will see that both descriptions yield qualitatively different behavior if one studies the dependency on parameters. This may serve as a warning: The relation between continuous and discrete–time models is generally not trivial at all.

4. Random Evolution

The descriptions by an ODE ($u' = f(u)$) or by a map ($\Phi : M \to M$) both lead to deterministic evolution. This just means that, in principle, all states ($u(t)$ or u_n) are determined by the law of evolution once an initial state at time zero, say, is fixed ($u(0) = u_0$). In the map case, determinism only holds for future states, i.e., for u_n with $n \geq 0$.

In contrast, what is random evolution? It turns out to be quite difficult to give a mathematically precise definition of *randomness*. To have a first glance at the difficulties, consider the sequence of digits

$$4252603764690804\ldots \tag{2.4}$$

which looks pretty random. The above sequence is obtained from the digits of

$$\pi = 3.141592653589793\ldots \tag{2.5}$$

by adding one to each digit of π, except that 9 is replaced by zero. One does not consider the digits of π to be random since one can write an algorithm, with finitely many statements, which produces the sequence of the digits of π. Of course, the execution of the algorithm never comes to an end. Accepting this concept of non–randomness, the sequence (2.4) is also not random. It appears, then, that our notion of randomness depends on our *knowledge*.

In this regard, there are similarities to probability. If I do not know when you were born then, for me, the probability that you were born on a Monday is about one sevenths. Actually, of course, you were either born on a Monday or not. Thus, the probability that you were born on a Monday is either one (if you were born on a Monday) or zero (if you were not born on a Monday). This is not to say that probability theory is mathematically vague. In fact, through the work of Emile Borel (1871–1956) and Andrey Kolmogorov (1903–1987), probability theory has obtained an axiomatic formulation which makes it mathematically completely sound. Kolmogorov's fundamental work on probability [9] was originally published in German in 1933. There are also mathematically rigorous approaches to randomness. An exposition on random sequences is given in [17].

For the time being, let us ignore these foundational questions and take a practical approach to random evolution. In Chapter 11 we will consider the Gambler's Ruin Problem, which is an example of a discrete–time Markov chain.

The general set–up for such a Markov chain is as follows: Let

$$M = \{e_1, e_2, \ldots, e_N\}$$

denote a finite set, the set of all possible states of the system that we consider. The random evolution takes place in M, in discrete time–steps. We follow the common tradition and denote the discrete time variable here by

$$t = 0, 1, 2, \ldots$$

and denote the state of the system at time t by X_t, which is an element of M. The random evolution is then specified by a so–called transition matrix P, which is a real, nonnegative, column stochastic $N \times N$ matrix :

$$P = \begin{pmatrix} p_{11} & \cdots & p_{1N} \\ \vdots & \ddots & \vdots \\ p_{N1} & \cdots & p_{N1} \end{pmatrix} \quad \text{with} \quad \sum_{i=1}^{N} p_{ij} = 1 \quad \text{for} \quad j = 1, \ldots, N \, . \tag{2.6}$$

The elements in each column of P sum up to one, which makes P *column stochastic.* The interpretation of the matrix element p_{ij} is the following: The number p_{ij} is the probability that — at time $t+1$ — the system will be in state $X_{t+1} = e_i$ under the assumption that — at time t — it was in state $X_t = e_j$. In a formula:

$$p_{ij} = prob\Big(X_{t+1} = e_i \Big| X_t = e_j\Big) \, . \tag{2.7}$$

Consider the following example:

$$M = \{1, 2\}, \quad P = \begin{pmatrix} 0.9 & 0.01 \\ 0.1 & 0.99 \end{pmatrix} \, . \tag{2.8}$$

Here the state space consists of two states, the numbers 1 and 2. If the system is in state $X_t = 1$ then it will be in state $X_{t+1} = 1$ with probability 0.9 and in state $X_{t+1} = 2$ with probability 0.1. The probabilities 0.9 and 0.1 are the entries in the first column of P. If the system is in state $X_t = 2$ then it will be in state $X_{t+1} = 1$ with probability 0.01 and in state $X_{t+1} = 2$ with probability 0.99.

In each time step, the probability to leave state 1 is $p_{21} = 0.1$, to leave state 2 is $p_{12} = 0.01$. Thus, the system has a higher probability to leave state 1 than to leave state 2. We expect, then, that in the long run the system will spend more time in state 2 than in state 1. We will see in Chapter 11 how to address this precisely.

Two other questions that one can ask:

1. Assume $X_0 = 1$. What is the probability that $X_{10} = 2$?

2. Assume $X_0 = 1$. What is the smallest N so that $X_N = 2$ with probability $\frac{1}{2}$ or larger?

In Chapter 11 we will learn how to answer such questions.

Here is an important point: To model a random evolution one specifies *probabilities* of the evolution. Consequently, one can also only answer *probabilistic* questions.

For systems more complicated than (2.8) it may be a good idea to use (many) numerical realizations of the random evolution. This can often be done quite effectively using random number generators. In Chapter 11 we will illustrate this using Matlab's *rand* command.

Conceptually more difficult than discrete time Markov chains are random evolutions in continuous time. As an example, we will consider a stochastic model of a growth process in Chapter 13. We will analyze the model using the *Forward Kolmogorov Equations* and will compare the analysis with results of numerical simulations.

5. Discussion

Randomness: Consider a sequence of digits

$$s = (d_1 d_2 d_3 \ldots) \quad \text{where} \quad d_j \in \{0, 1, 2, \ldots, 9\} \ . \tag{2.9}$$

When should one call such a sequence *random*? Generalizing what we said about the sequences (2.4) and (2.5), we may attempt to get a definition as follows: If there exists a *finite* set of rules which describes the sequence (2.9) exactly, then we call the sequence *non–random*. If no such finite set of rules exists, then we call the sequence *random*.

Let us now discuss and criticize this definition of randomness and non–randomness of sequences (2.9). The set of finite rules (or finite algorithms) is denumerable, i.e., its cardinality equals that of $\mathbb{N}$, the set of positive integers. (See the definition of *cardinality* below where we give Cantor's diagonalization argument.) Therefore, the set of all non–random sequences is also denumerable.

Now consider a sequence like

$$1133778844\ldots \tag{2.10}$$

where

$$d_j = d_{j+1} \quad \text{for all odd } j \ . \tag{2.11}$$

Using Cantor's diagonalization argument (see below), it is not difficult to show that the set of all sequences satisfying (2.11) is larger than denumerable. Therefore, most of theses sequences must be random, according to the definition that we attempt. However, if (2.11) holds for all j, then the corresponding sequence obeys a *local law of non–randomness.* Is it justified to call such a sequence random?

It is clear that a satisfactory mathematical definition of the randomness of a sequence is not easy.

Remark on Randomness and Creation vs. Evolution: Charles Darwin's (1809–1882) theory of evolution [4] says, very roughly, that mutations occur *randomly*, but then

[4] In the remainder of the book, the word evolution has a broader meaning: the change over time.

the process of natural selection (survival of the fittest) sets in. Together, the two processes, random mutations and selection, led to the great variety of lifeforms that we see today, well adapted to their environment. What does *randomness* of the mutations mean? One should note that randomness of any process is not an objective property of the process. Because we lack knowledge, we can only *describe* the occurrence of mutations as random. We just don't know, for example, when and where any gamma ray interacts with the DNA of any animal or plant. We will never have this kind of knowledge. However, for an intelligence that knows *everything*, nothing is random. Maybe evolution was and is the way to create.

History: In this short chapter, the Russian mathematician Andrey Kolmogorov (1903–1987) was mentioned twice: He laid the foundations to axiomatic probability theory [9] and developed important equations in the field of stochastic processes. Furthermore, Kolmogorov is well known for his work in turbulence theory and for the Kolmogorov–Arnold–Moser Theorem, an important perturbation result for Hamiltonian systems of classical mechanics.

In addition, Kolmogorov was the founder of algorithmic complexity theory, also known as Kolmogorov complexity. Roughly, if one has a finite sequence of digits

$$s = (d_1 d_2 \dots d_N)$$

then its Kolmogorov complexity $d(s)$ is the size of the smallest code which produces this string. If $d(s)$ is much smaller than the length N of the string s, then s is non–random. Kolmogorov complexity gives a *quantitative* approach to randomness and non–randomness.

A Markov chain, which was introduced in the previous section, is named after the Russian mathematician Andrey Markov (1856–1922). An important property of such a chain is expressed in equation (2.7). Here one should note that the probability distribution of the state X_{t+1} depends *only* on the state X_t and does not depend on the history $X_0, X_1, \dots, X_{t-1}$. This so–called Markov property is a realistic assumption for many random evolutions and simplifies their analysis. The Gambler's Ruin Problem in Chapters 11 and 12 will illustrate this.

Cantor's Diagonalization Argument: George Cantor (1845–1918), a German mathematician, was the founder of set theory. According to Cantor's definition, two set A and B have the same cardinality if there is a map $f : A \to B$ which is one–to–one and onto. The set $\mathbb{N}$ of positive integers and the set of all real numbers x with $0 \leq x \leq 1$ are both infinite, but do not have the same cardinality. Cantor gave a beautiful argument, known has Cantor's diagonalization argument: Suppose the set $[0, 1]$ has the same cardinality as $\mathbb{N}$. Then one can write all $x \in [0, 1]$ as a sequence,

$$\begin{aligned} x_1 &= 0.d_{11}d_{12}d_{13}\ldots \\ x_2 &= 0.d_{21}d_{22}d_{23}\ldots \\ x_3 &= 0.d_{31}d_{32}d_{33}\ldots \\ &\vdots \end{aligned}$$

where the $d_{nj} \in \{0,1,2,\ldots,9\}$ are the decimal digits of x_n. Now consider the number given by the diagonal entries,

$$y = 0.d_{11}d_{22}d_{33}\ldots$$

and change every digit. For example, let

$$\tilde{d}_{jj} = \begin{cases} d_{jj}+1 & \text{for } d_{jj} < 9 \\ 0 & \text{for } d_{jj} = 9 \end{cases}$$

The number

$$\tilde{y} = 0.\tilde{d}_{11}\tilde{d}_{22}\tilde{d}_{33}\ldots$$

lies in $[0,1]$, but $\tilde{y}$ does not show up in the above list of the x_n, because the n–th digit of x_n is d_{nn} and the n–th digit of $\tilde{y}$ is $\tilde{d}_{nn}$, which is different from d_{nn}. This contradiction shows that the set $[0,1]$ is not denumerable.

Chapter 3

Deterministic Systems: Outline of Advanced Topics

Summary: In this chapter we outline some topics for deterministic systems. The systems are either described by an initial value problem for an ODE:

$$u' = f(u), \quad u(0) = u_0 \ , \tag{3.1}$$

or by a map $\Phi : M \to M$, which leads to the discrete–time dynamics:

$$u_{n+1} = \Phi(u_n), \quad n = 0, 1, \ldots \tag{3.2}$$

The topics which we outline are: Determinism for planetary motion (Newton and Bessel), time reversibility, sensitive dependence on initial conditions and Lyapunov exponents, bifurcations, variation on different time scales.

1. Planetary Motion: Example for Determinism

Newton's 2nd law *(force = mass times acceleration)* together with the inverse square law of gravitational attraction can be used to formulate the motion of two point masses as a dynamical system (3.1). The vector $u(t)$ will have twelve components, the six position and six velocity coordinates of the two masses. It is rather simple, and can be physically motivated, to reduce the problem from twelve time dependent variables to only two. But then it is not easy to show that the resulting orbit is an ellipse, a parabola, or a hyperbola of one mass around the other. In Newton's Principia this derivation was carried out with arguments of a more geometric nature than we use today. The derivation which we give in Chapter 4 uses calculus instead of geometry.

Even if one knows that the orbit is an ellipse, say, it is far from trivial to obtain an *explicit* expression for the position vector as a function of time t. In one way or another, one has to solve Kepler's equation,

$$\frac{2\pi t}{T} = \alpha - \varepsilon \sin \alpha \ ,$$

for $\alpha = \alpha(t)$. In Kepler's equation, T is the period of the motion and ε is the eccentricity of the ellipse. Kepler's equation can be solved *approximately* for the unknown α by iteration, but this does not amount to an explicit solution. Interestingly, this difficulty lead the astronomer Friedrich Wilhelm Bessel (1784–1846) to introduce a new class of functions, now called Bessel functions, which play an important role in mathematical physics and engineering.

In Chapter 4 we carry out the details. The math is not easy and the details may be skipped at a first reading. They will not be used later.

2. Reversibility in Time

The equations of planetary motion give an example of a system which is reversible in time. If we change the signs of all initial velocities, the planets will trace out the same orbits, but backwards. There are reasons to believe that, in some generalized sense, all *fundamental* laws of physics are time reversible. However, the observed macro world does not show time reversibility at all. Coffee spills onto the carpet, but never jumps back into the cup.

If the dynamical laws, as expressed by $u' = f(u)$, are time reversible, but one observes a *direction* or an *arrow* of time, then one can try to argue that the arrow is a consequence of the (special) initial condition $u(0) = u_0$. We may then speculate that, ultimately, the very special initial condition of the *big bang* enables us to spill our coffee.

In Chapter 5 we will formalize the mathematical notion of time reversibility.

In thermodynamics, the *irreversibility* of time is important. In various ways, it can be expressed as the *second law of thermodynamics*. First insights into this law were obtained by Sadi Carnot (1796–1832) in the early 18 hundreds. He studied how to transform heat (random motion on the micro scale) into mechanical work (deterministic motion on the macro scale). Carnot's idealized heat engine is the subject of Section 4..

In the last chapter of the book we illustrate time irreversibility by a simple model of random evolution. The process quickly leads from order to chaos, but going back to order turns out to be extremely unlikely. This is a probabilistic view of irreversibility in time.

3. Sensitive Dependence on Initial Conditions

Though the evolution determined by the systems (3.1) and (3.2) is deterministic, [1] it may be practically impossible to predict future states accurately. One reason is that the initial

[1] We always assume that the function $f(u)$ is smooth so that the initial value problem (3.1) is uniquely solvable.

state u_0 is typically not exactly known and an error in u_0 may grow exponentially in time. One can measure this growth by a Lyapunov exponent [2] .

Let us explain this concept in the simple case where M is a subinterval of the real line and $\Phi : M \to M$ is a smooth map. Consider an orbit $\{u_n\}_{n=0,1,\ldots}$ determined by the iterative process

$$u_{n+1} = \Phi(u_n), \quad n = 0, 1, \ldots$$

where $u_0 \in M$ is some fixed initial value. What happens if we perturb u_0 and start the orbit at $v_0 = u_0 + \varepsilon$? Let $v_n = \Phi^n(v_0)$ denote [3] the corresponding orbit starting at v_0.

By Taylor's formula :

$$\begin{aligned} v_1 &= \Phi(v_0) \\ &= \Phi(u_0 + \varepsilon) \\ &= \Phi(u_0) + \Phi'(u_0)\varepsilon + \mathcal{O}(\varepsilon^2) \\ &= u_1 + \Phi'(u_0)\varepsilon + \mathcal{O}(\varepsilon^2) \end{aligned}$$

Thus, after one time step, the initial error $\varepsilon = v_0 - u_0$ becomes

$$v_1 - u_1 = \Phi'(u_0)\varepsilon + \mathcal{O}(\varepsilon^2) .$$

If we neglect the $\mathcal{O}(\varepsilon^2)$–term we see that the initial error ε is multiplied by $\Phi'(u_0)$ after one time step. It is not difficult to generalize this to n time steps. One finds that

$$v_n - u_n = (\Phi^n)'(u_0)\varepsilon + \mathcal{O}(\varepsilon^2) .$$

Here the $\mathcal{O}(\varepsilon^2)$–term depends on n. By the chain rule,

$$(\Phi^2)'(u_0) = \Phi'(\Phi(u_0))\Phi'(u_0) = \Phi'(u_0)\Phi'(u_1)$$

and, by induction,

$$(\Phi^n)'(u_0) = \Phi'(u_0)\Phi'(u_1)\cdots\Phi'(u_{n-1}) .$$

Thus, the linearized error propagator after n steps is

$$\begin{aligned} \lim_{\varepsilon\to 0} \frac{1}{\varepsilon}\Big(\Phi^n(u_0+\varepsilon) - \Phi^n(u_0)\Big) &= (\Phi^n)'(u_0) \\ &= \Phi'(u_0)\Phi'(u_1)\cdots\Phi'(u_{n-1}) . \end{aligned}$$

[2] named after Aleksandr Mikhailovich Lyapunov (1857–1918)

[3] Here the map $\Phi^n = \Phi\circ\Phi\circ\ldots\circ\Phi$ is obtained by n applications of Φ; for example, $\Phi^2(u) = (\Phi\circ\Phi)(u) = \Phi(\Phi(u))$ for all arguments u.

Note that we must multiply the local error propagators $\Phi'(u_j)$ along the orbit.

If we expect an (approximate) exponential behavior of the error $|v_n - u_n|$ with increasing n we may try to write

$$|(\Phi^n)'(u_0)| \sim e^{\lambda n}$$

where λ is a parameter that needs to be determined. Taking logarithms, one finds that

$$\lambda \sim \frac{1}{n} \log |(\Phi^n)'(u_0)| = \frac{1}{n} \sum_{j=0}^{n-1} \log |\Phi'(u_j)| \; .$$

This motivates the following definition.

Definition 3.1: *If the following limit exists*

$$\lambda := \lim_{n \to \infty} \frac{1}{n} \sum_{j=0}^{n-1} \log |\Phi'(u_j)| \tag{3.3}$$

then λ is called the Lyapunov exponent of the map Φ along the orbit $\{u_j\}_{j=0,1,\ldots}$.

For generalizations and applications to differential equations see, for example, [1].

A positive Lyapunov exponent λ implies exponential growth of any small initial error with increasing time. This often limits predictability in practice.

As a simple example, consider the so–called Bernoulli shift.

Example: (Bernoulli Shift)

$$M = [0,1), \quad \Phi(u) = \begin{cases} 2u & \text{for} \quad 0 \le u < \frac{1}{2} \; , \\ 2u - 1 & \text{for} \quad \frac{1}{2} \le u < 1 \; . \end{cases} \tag{3.4}$$

This map Φ is not differentiable at $u = \frac{1}{2}$, but we can apply the above considerations as long as none of the elements of the orbit $\{u_n\}_{n=0,1,\ldots}$ equals $\frac{1}{2}$. One then obtains that

$$\Phi'(u_j) = 2 \quad \text{for all} \quad j = 0, 1, \ldots$$

thus

$$\lambda = \log 2 > 0 \; .$$

In this example, any small initial error ε grows like

$$e^{\lambda n} \varepsilon = 2^n \varepsilon \; ,$$

with increasing time, i.e., the error doubles in each time step. The dynamics of the Bernoulli shift is quite interesting and one can analyze it rather completely by writing the numbers u_n in base two. We will carry this out in Chapter 6.

4. Averages

If a system has a positive Lyapunov exponent then the accurate long time prediction for *individual* trajectories is often practically impossible. Nevertheless, the *average* long–time behavior of the system may be robustly determined. In Chapter 6 we will illustrate this using the dynamics determined by the logistic map

$$\Phi_{logistic}(u) = 4u(1-u), \quad M = [0,1] ,$$

as an example. There is a simple relation between the logistic map and the Bernoulli shift, which implies sensitive dependence on initial conditions for the logistic map.

A different and historically very important example for well–determined average behavior is given by Maxwell's velocity distribution of particles in a gas. Maxwell's beautiful arguments, based on symmetry assumptions of the distribution function, will be given in Section 3..

5. Dependence on Parameters

In applications, the functions $f(u)$ and $\Phi(u)$, which determine the evolution in (3.1) and (3.2), will often depend on parameters. Let us restrict ourselves to the case where f or Φ depends on one real parameter λ and write $f = f(u,\lambda)$ and $\Phi = \Phi(u,\lambda)$. We may think of λ as a control parameter that can be changed by an experimenter. To have a difficult example, we may think of λ as the CO_2 concentration in the atmosphere, which has been increased from preindustrial times by about 30 percent.

Small changes in the parameter λ will typically lead to (quantitatively) small changes in the solutions of the systems $u' = f(u,\lambda)$ and $u_{n+1} = \Phi(u_n,\lambda)$.

If there is a *qualitative* change of the behavior of the solution at some parameter value $\lambda = \lambda_0$, then one calls λ_0 a bifurcation value.

For example, a fixed point $u(\lambda)$ of the evolution may be stable for $\lambda < \lambda_0$, but become unstable for $\lambda > \lambda_0$. Then, if λ changes from $\lambda < \lambda_0$ to $\lambda > \lambda_0$, a *qualitative* change of the dynamics will occur near the fixed point: We have a bifurcation at $\lambda = \lambda_0$.

An important, but often difficult question of bifurcation theory is the following: Suppose some phenomenon is stable (and therefore observable) for $\lambda < \lambda_0$, but unstable for $\lambda > \lambda_0$. Which (new) phenomenon will take over stability for $\lambda > \lambda_0$? In this text we will address this question only using very simple examples; see Chapter 8. In this context, we will also discuss the hysteresis phenomenon.

6. Variation on Different Time Scales

A simple example of an initial value problem whose solution varies on two time scales is

$$u' = \frac{i}{\varepsilon}(u - \sin t), \quad u(0) = 1 , \tag{3.5}$$

where $0 < \varepsilon << 1$. Here i is complex and $i^2 = -1$. In an ODE course you may have learned how to solve (3.5) exactly. We will ignore the exact solution formula, however. Instead, we will derive an approximate solution. The process gives insight into the two different time scales that are present.

Let us first note that the *homogeneous* equation corresponding to the differential equation (3.5) is $u' = \frac{i}{\varepsilon}u$. Its solution

$$u_{hom}(t) = e^{it/\varepsilon} = \cos(t/\varepsilon) + i\sin(t/\varepsilon)$$

varies fast. For example, if $\varepsilon = \frac{1}{50}$ and $0 \le t \le 2\pi$, then $u_{hom}(t)$ goes through 50 full oscillations. On the other hand, the forcing term in (3.5) is

$$-\frac{i}{\varepsilon}\sin t$$

which goes through only one oscillation for $0 \le t \le 2\pi$. One says that $u_{hom}(t) = e^{it/\varepsilon}$ varies on the fast time scale whereas the above forcing term varies on the slow time scale.

What about the solution $u(t)$ of (3.5)? We want to argue that the solution varies on both scales, the slow and the fast. Multiplying the differential equation by ε and then formally setting $\varepsilon = 0$ yields the approximate solution

$$u_0(t) = \sin t \ .$$

This function does not satisfy the initial condition $u(0) = 1$, however. Adding the solution $u_{hom}(t) = e^{it/\varepsilon}$ of the homogeneous equation to $u_0(t)$ one obtains

$$u_1(t) = e^{it/\varepsilon} + \sin t \ ,$$

a function which satisfies the initial condition and satisfies the ODE

$$\varepsilon u' = i(u - \sin t)$$

up to order ε. This suggests that $u_1(t)$ agrees with the exact solution of (3.5) up to a term which is $\mathcal{O}(\varepsilon)$. In Chapter 7 we will show that this is true. A plot of the real part,

$$\operatorname{Re} u_1(t) = \cos(t/\varepsilon) + \sin t, \quad 0 \le t \le 2\pi \ ,$$

for $\varepsilon = 0.02$ is shown in Figure 3.1.

The figure also shows the slowly varying part, $u_0(t) = \sin t$. The function $u_1(t)$ is a simple example of a function varying on two time scales, the slow scale t and the fast scale t/ε.

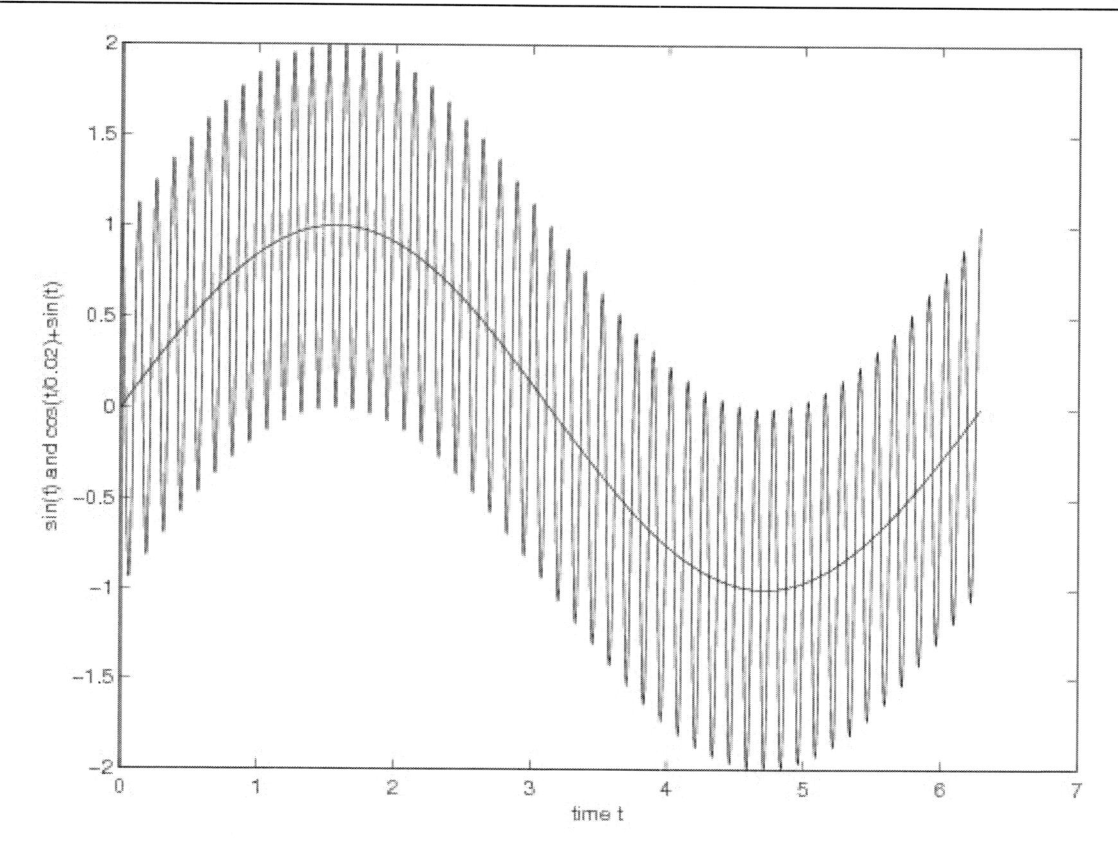

Figure 3.1. Fast and slowly varying functions

In applications, one often deals with phenomena that vary on different time and space scales. An important, but very difficult, example is the weather/climate system. The weather changes on the fast time scale, but for climate predictions we are interested in changes on the slow time scale. If one has good equations to model a phenomenon on the fast time scale, how can one get equations describing the evolution of slowly varying parts of the solution? In the simple example (3.5), the process was to multiply the differential equation by ε, then setting $\varepsilon = 0$ and ignoring the initial condition. Remarkably, this idea often works if one can identify a small parameter ε in the equations. We refer to the literature for many extensions; see, for example, [10].

Related to the above is the following: For a gas we have two models, the micro model of many molecules bouncing around at the fast time scale and the macro model of slowly varying variables like temperature and pressure. To connect these two descriptions is the subject of kinetic theory.

Imagine a container with 2 grams of hydrogen. It contains [4] $N = 6.02 * 10^{23}$ H_2–molecules, which bounce around at the fast time scale. This is the micro picture. In the macro model we would like to understand how macro variables like the temperature T, the pressure p, and the volume V of the gas are related to each other. For example, if we change the volume at constant temperature, how does the pressure change? Clearly, to compute this directly using the micro model of $N = 6.02 * 10^{23}$ bouncing particles would be extremely difficult. We much rather work directly with the macro variables T, p, V. Luckily, thermodynamics provides us with the relevant equations and (except for extreme situations) we can ignore the micro picture. It is, of course, still interesting and challenging to connect the micro and macro descriptions. Here a first contribution was made by Daniel Bernoulli in 1738. Bernoulli's aim was to derive Boyle' law ($pV = const$ at constant temperature) from a bouncing ball picture of the gas. In Section 2. we give Bernoulli's arguments.

[4] The number $N = 6.02 * 10^{23}$ is known as Avogadro's or Loschmidt's number . It gives the number of molecules in one mole of a substance.

Chapter 4

Planetary Motion

Summary: Based on observations by the Danish astronomer Tycho Brahe (1546–1601), Johannes Kepler (1571–1630) formulated his three laws of planetary motion:

1. The orbit of every planet is an ellipse with the sun at a focus.
2. The line from the sun to a planet sweeps out equal areas in equal time intervals.
3. If T denotes the orbital period of a planet and a denotes the major semi–axis of its orbit, then T^2/a^3 is a constant independent of the planet.

Kepler's three laws were explained by Isaac Newton (1643–1727) using Newton's second law ($force = mass * acceleration$) and a central gravitational field which decays like $const/r^2$. Newton's demonstration, published in 1687 in his monograph *Philosophiae Naturalis Principia Mathematica*, was very important for the history of science. Not only did it remove the last doubts about the heliocentric system (already suggested by Nicolaus Copernicus (1473–1543)), but it greatly advanced the Scientific Revolution. One may say that, because of Newton's success, classical mechanics became a model for all sciences.

In this chapter we use Newton's second law and the law of gravitational attraction to formulate the motion of a planet around the sun as a system of ODEs. We then use this formulation to deduce Kepler's laws. To get the explicit time dependence of a planet in its orbit requires solving a transcendental equation, called Kepler's equation. More than one hundred years after Newton, the astronomer F.W. Bessel introduced a new class of functions, now called Bessel functions, to attack this problem. This is our second subject in this chapter.

The equations for planetary motion are deterministic, and Newton's success contributed to a deterministic world view. Does the whole universe evolve deterministically? Is free will an illusion? We end the chapter with a short discussion.

1. Outline

Because of their historical importance as an example for determinism, we derive here Kepler's laws in some detail. This is carried out in Sections 2. – 6.. To obtain the position of a planet as an explicit function of time, requires in addition solving Kepler's equation

$$\frac{2\pi t}{T} = \alpha - \varepsilon \sin \alpha \tag{4.1}$$

for $\alpha = \alpha(t)$. This problem led the astronomer Friedrich Wilhelm Bessel (1784–1846) to introduce a new class of functions, now called Bessel functions, which to this date play an important role in mathematics and engineering. We give the details in Sections 7.–8.. Though the mathematics in this chapter is quite classical, it is by no means simple.

As a starting point, we consider two point masses under their mutual gravitational attraction and formulate the law of their motion as an initial value problem. In principle, Kepler's laws then follow from Newton's second law ($force = mass * acceleration$) and the inverse–square law of gravitational attraction. A rough outline is as follows: Formulated as a first order system, the two–body problem contains twelve time dependent variables, the six position coordinates and the six velocity components of the two bodies (modeled as points). Going into the coordinate frame where the center of mass rests, removes six variables. Equivalently, one obtains the one–body problem. The motion takes place in the plane orthogonal to the constant vector of angular momentum, reducing from six variables to four. Constancy of the energy and the size of the angular momentum leads to a system of two variables, $r(t)$ and $\theta(t)$. See (4.7), (4.8). This system is difficult to solve, but one can eliminate time and obtain an equation for

$$\frac{dr}{d\theta} = \frac{dr/dt}{d\theta/dt} \; .$$

The solution of this equation finally leads to Kepler's ellipse. Explicit solutions for $r(t)$ and $\theta(t)$ can be obtained if, in addition, one solves Kepler's equation (4.1).

2. The Two Body Problem: Reduction to One Body in a Central Field

Denote the masses by m_1 (sun) and m_2 (planet) and denote the gravitational constant by G. The position vectors of the bodies are $\mathbf{r}_1$ and $\mathbf{r}_2$. Further, let[1]

$$\mathbf{r} = \mathbf{r}_2 - \mathbf{r}_1, \quad r = |\mathbf{r}| \; .$$

[1]We denote the Euclidean inner–product of two vectors $\mathbf{a}$ and $\mathbf{b}$ by $\langle \mathbf{a}, \mathbf{b} \rangle$, and $|\mathbf{a}| = \sqrt{\langle \mathbf{a}, \mathbf{a} \rangle}$ denotes the Euclidean length.

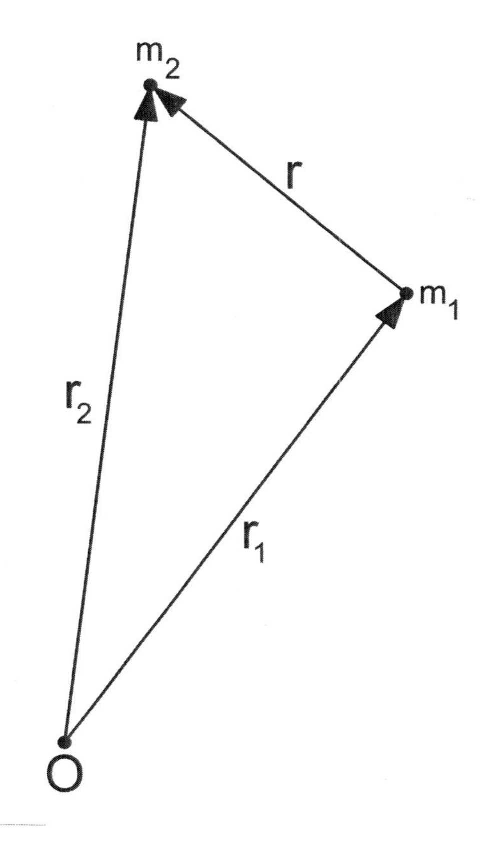

Figure 4.1. Two point masses

The vector $\mathbf{r}$ is directed from m_1 to m_2, i.e., from the sun to the planet. By the law of gravitational attraction: [2]

$$m_1\ddot{\mathbf{r}}_1 = Gm_1m_2\frac{\mathbf{r}}{r^3} =: \mathbf{F} \tag{4.2}$$

$$m_2\ddot{\mathbf{r}}_2 = -\mathbf{F} \tag{4.3}$$

Set $M = m_1 + m_2$. Then

$$\mathbf{R} = \frac{1}{M}(m_1\mathbf{r}_1 + m_2\mathbf{r}_2)$$

is the position vector of the center of mass. From (4.2) and (4.3):

[2] In this chapter we denote the time derivative of a function $u(t)$ by $\dot{u}(t)$ instead of $u'(t) = \frac{du}{dt}(t)$. The dot notation goes back to Isaac Newton.

$$\ddot{\mathbf{R}} = 0 \; .$$

By choosing an inertial frame in which the center of mass rests at the origin, we may assume that $\mathbf{R}(t) \equiv 0$. (This amounts to a suitable restriction of the initial data.)

We have

$$m_1 m_2 \ddot{\mathbf{r}}_1 = m_2 \mathbf{F} \quad \text{and} \quad m_1 m_2 \ddot{\mathbf{r}}_2 = -m_1 \mathbf{F} \; .$$

Therefore,

$$m_1 m_2 \ddot{\mathbf{r}} = -M\mathbf{F} \; .$$

Dividing by $M = m_1 + m_2$ we obtain

$$m\ddot{\mathbf{r}} = -k \, \frac{\mathbf{r}}{r^3}, \quad k = G m_1 m_2$$

with

$$m = \frac{m_1 m_2}{m_1 + m_2} \; .$$

3. One Body in a Central Field

The above equation

$$m\ddot{\mathbf{r}} = -k \, \frac{\mathbf{r}}{r^3} \tag{4.4}$$

describes the motion of a body in an attracting central force field which decays like r^{-2}.

Angular Momentum and Kepler's Second Law. The vector

$$\mathbf{L} = \mathbf{r} \times m\dot{\mathbf{r}}$$

is the angular momentum. Since $\ddot{\mathbf{r}}$ has the same direction as $\mathbf{r}$ it follows that $\dot{\mathbf{L}} = 0$, thus

$$\mathbf{L}(t) = \mathbf{L}_0$$

is a constant vector, not changing in time. In terms of the initial data

$$\mathbf{r}(0) = \mathbf{r}_0, \quad \dot{\mathbf{r}}(0) = \dot{\mathbf{r}}_0$$

we have

$$\mathbf{L}(t) \equiv \mathbf{L}_0 = \mathbf{r}_0 \times m\dot{\mathbf{r}}_0 \; .$$

We assume now that $\mathbf{L}_0$ is a non–zero vector of length

$$|\mathbf{L}_0| = l \ .$$

(One can also consider the case $\mathbf{L}_0 = 0$. This leads to motion on a straight line and possible collision of the body with the center.)

Since $\mathbf{r}(t)$ is orthogonal to $\mathbf{L}_0$ at all time, the motion takes place in the plane through the origin orthogonal to $\mathbf{L}_0$.

We choose coordinates so that

$$\mathbf{r} = r(c, s, 0) \quad \text{with} \quad c = \cos\theta, \quad s = \sin\theta \ . \tag{4.5}$$

Thus, (r, θ) are polar coordinates in the plane orthogonal to $\mathbf{L}_0$. It then follows that

$$\mathbf{L} = \mathbf{r} \times m\dot{\mathbf{r}} = mr^2\dot{\theta}\mathbf{e}_3 \quad \text{where} \quad \mathbf{e}_3 = (0, 0, 1) \ .$$

Therefore,

$$|\mathbf{L}(t)| = |\mathbf{L}_0| = l = mr^2\dot{\theta} \ . \tag{4.6}$$

The geometric interpretation of the time–independence of

$$\frac{1}{2}r^2\frac{d\theta}{dt}$$

is Kepler's second law.

Conservation of Energy. If we define

$$\begin{aligned} E_{kin} &:= \frac{m}{2}|\dot{\mathbf{r}}|^2 \\ E_{pot} &:= -\frac{k}{r} \end{aligned}$$

and

$$E := E_{kin} + E_{pot}$$

then

$$\begin{aligned} \dot{E} &= m\langle \dot{\mathbf{r}}, \ddot{\mathbf{r}} \rangle + k\dot{r}/r^2 \\ &= -k\langle \dot{\mathbf{r}}, \mathbf{r}/r^3 \rangle + k\dot{r}/r^2 \\ &= 0 \end{aligned}$$

In the last equation we have used that $\langle \dot{\mathbf{r}}, \mathbf{r} \rangle = \dot{r}r$, which follows from the form (4.5).

Since $\dot{E} = 0$, the energy is constant in time:

$$E(t) = E_0$$

is a constant determined by the initial condition. Below we will assume that $E_0 < 0$. The cases $E_0 = 0$ and $E_0 > 0$ can be treated similarly, leading to a parabolic and a hyperbolic orbit, respectively. Using the equation $\mathbf{r} = r(c, s, 0)$ one finds that

$$|\dot{\mathbf{r}}|^2 = \dot{r}^2 + r^2\dot{\theta}^2 \; .$$

Therefore,

$$\frac{m}{2}(\dot{r}^2 + r^2\dot{\theta}^2) - \frac{k}{r} = E_0 \; .$$

From (4.6):

$$mr^2\dot{\theta}^2 = \frac{l^2}{mr^2}$$

To summarize, constancy of energy ($E(t) = E_0$) and the size of angular momentum ($|\mathbf{L}(t)| = l$) lead to the fundamental relation

$$E_0 = \frac{m}{2}\dot{r}^2 + \frac{l^2}{2mr^2} - \frac{k}{r} \; ,$$

which is a differential equation for $r(t)$. From Newton's 2nd law and the law of gravity we have derived the dynamical equations

$$\frac{dr}{dt} = \pm\Big(\frac{2E_0}{m} + \frac{2k}{mr} - \frac{l^2}{m^2r^2}\Big)^{1/2} \tag{4.7}$$

$$\frac{d\theta}{dt} = \frac{l}{mr^2} \tag{4.8}$$

Here (r, θ) are polar coordinates in the plane perpendicular to the (constant) angular momentum. The constants E_0 and l are determined by the initial data $\mathbf{r}_0$ and $\dot{\mathbf{r}}_0$.

4. The Equation for an Ellipse in Polar Coordinates

Cartesian Coordinates. For the following, see Figure 4.2. Let $0 < c < a$ and let

$$F_1 = (c, 0), \quad F_2 = (-c, 0) \; .$$

Consider all points $P = (x, y)$ with

$$|F_1 - P| + |F_2 - P| = 2a \; .$$

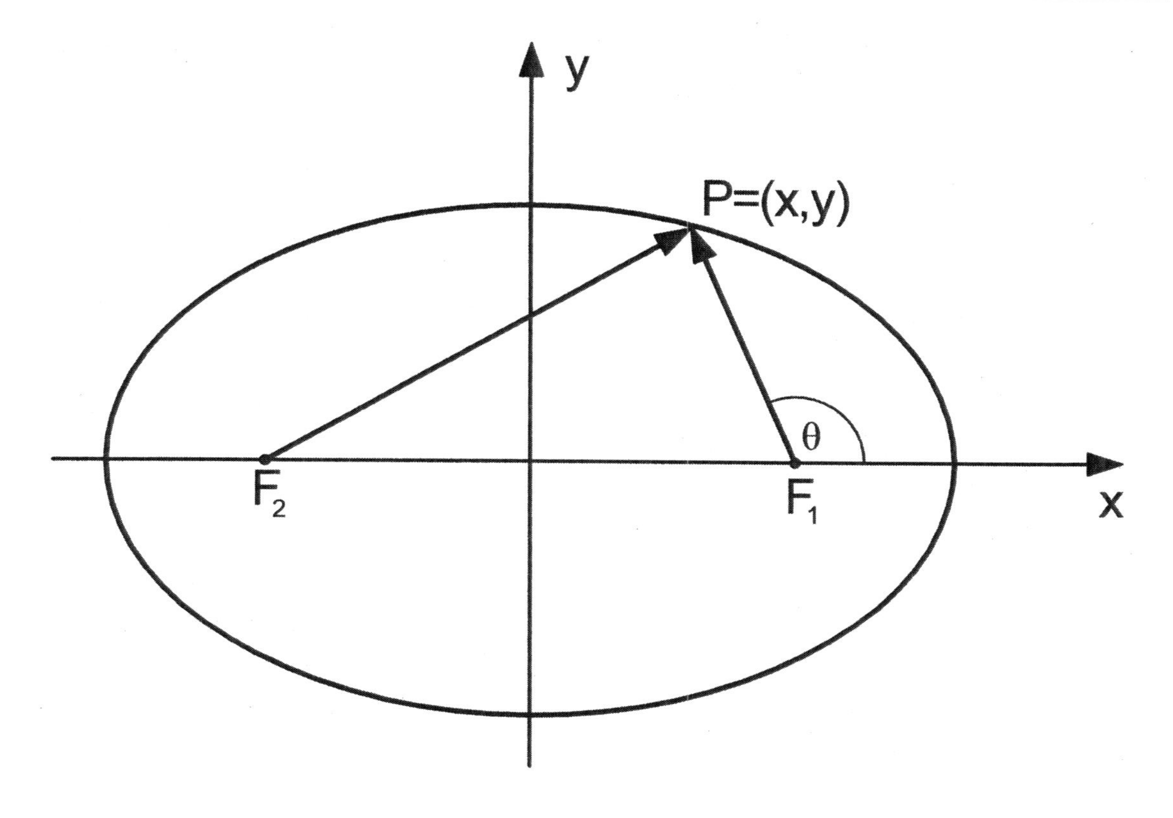

Figure 4.2. Ellipse with foci F_1 and F_2

By definition, these points $P = (x, y)$ form the ellipse with foci F_1, F_2 and major semi–axis a. We want to derive the equation satisfied by (x, y). To this end, let

$$\begin{aligned} d_1^2 &= |F_1 - P|^2 = (x-c)^2 + y^2 \\ d_2^2 &= |F_2 - P|^2 = (x+c)^2 + y^2 \end{aligned}$$

The equation

$$d_1 + d_2 = 2a$$

is equivalent to

$$d_2^2 = (2a - d_1)^2 \ ,$$

thus

$$(x+c)^2 + y^2 = 4a^2 - 4ad_1 + (x-c)^2 + y^2 \ .$$

This is equivalent to

$$4cx - 4a^2 = -4ad_1 \ ,$$

or

$$a^2 - cx = ad_1 \ .$$

Squaring yields

$$a^4 - 2a^2cx + c^2x^2 = a^2(x^2 - 2cx + c^2 + y^2)$$

or

$$a^4 + c^2x^2 = a^2x^2 + a^2c^2 + a^2y^2 \ .$$

One obtains the equivalent condition

$$a^2(a^2 - c^2) = x^2(a^2 - c^2) + a^2y^2 \ .$$

If we define $b > 0$ by $b^2 = a^2 - c^2$ then we obtain

$$1 = \left(\frac{x}{a}\right)^2 + \left(\frac{y}{b}\right)^2 \ .$$

This is the well–known equation for an ellipse with semi–axes a and b in Cartesian coordinates.

Polar Coordinates. Denote polar coordinates centered at F_1 by (r, θ) and let $P = (x, y)$ denote a point on the ellipse with polar coordinates (r, θ). The distances of P to the foci are

$$d_1 = |F_1 - P| = r, \quad d_2 = |F_2 - P| = 2a - r \ .$$

If $\phi = \pi - \theta$ then the cosine theorem in the triangle F_1PF_2 yields

$$(2a - r)^2 = 4c^2 + r^2 - 4cr\cos\phi \ .$$

Therefore,

$$4a^2 - 4ar + r^2 = 4c^2 + r^2 - 4cr\cos\phi \ .$$

This yields

$$a^2 - c^2 = r(a - c\cos\phi) \ .$$

Introduce the eccentricity $\varepsilon = c/a$ of the ellipse to obtain

$$r = \frac{a(1 - \varepsilon^2)}{1 + \varepsilon\cos\theta} \ . \tag{4.9}$$

This is the equation of an ellipse in polar coordinates, and we are now prepared to derive Kepler's first law.

For later reference, note that substituting $\theta = 0$ in equation (4.9) yields

$$r(0) = a(1-\varepsilon) = a - c\,. \tag{4.10}$$

5. The Kepler Orbit

Recall the dynamical equations (4.7), (4.8), which we derived above. We repeat them here:

$$\begin{aligned}\frac{dr}{dt} &= \pm\Big(\frac{2E_0}{m} + \frac{2k}{mr} - \frac{l^2}{m^2r^2}\Big)^{1/2}\\ \frac{d\theta}{dt} &= \frac{l}{mr^2}\end{aligned}$$

The equations yield [3]

$$\frac{dr}{d\theta} = \pm r^2\Big(A + \frac{B}{r} - \frac{1}{r^2}\Big)^{1/2} \tag{4.11}$$

with

$$\begin{aligned}A &= \frac{2E_0}{m}\cdot\frac{m^2}{l^2} = \frac{2mE_0}{l^2}\\ B &= \frac{2k}{m}\cdot\frac{m^2}{l^2} = \frac{2mk}{l^2}\end{aligned}$$

We will solve the differential equation (4.11) for $r = r(\theta)$ using separation of variables. Formally:

$$\frac{dr}{\pm r^2\Big(A + \frac{B}{r} - \frac{1}{r^2}\Big)^{1/2}} = d\theta\,.$$

In the indefinite integral

$$Int = \int \frac{dr}{\pm r^2\Big(A + \frac{B}{r} - \frac{1}{r^2}\Big)^{1/2}}$$

[3]Equation (4.11) is obtained by formally taking the quotient of the two dynamical equations, which amounts to an elimination of time. It would be more rigorous, but cumbersome, to introduce a new variable $\tilde{r}(\theta) = r(t)$ if $\theta = \theta(t)$. As is usually the case, what is suggested by Leibniz' notation turns out to be correct.

use the substitution

$$r = \frac{1}{u}, \quad dr = -\frac{du}{u^2}, \quad -\frac{dr}{r^2} = du$$

to obtain

$$Int = \int \pm(A + Bu - u^2)^{-1/2}\, du, \quad u = 1/r\ .$$

The integral can be evaluated in terms of arccos. Note that the identity

$$\alpha = \arccos(\cos\alpha)$$

yields

$$\arccos'(y) = \pm(1 - y^2)^{-1/2}\ .$$

Set

$$h(u) = \arccos\Big(\frac{2u - B}{q}\Big)$$

where the constant q is to be determined. We have

$$\begin{aligned} h'(u) &= \pm\frac{2}{q}\Big(1 - q^{-2}(4u^2 - 4uB + B^2)\Big)^{-1/2} \\ &= \pm 2\Big(q^2 - 4u^2 + 4uB - B^2\Big)^{-1/2} \\ &= \pm\Big(\frac{1}{4}(q^2 - B^2) + uB - u^2\Big)^{-1/2} \end{aligned}$$

If

$$\frac{1}{4}(q^2 - B^2) = A$$

then

$$h'(u) = \pm(A + Bu - u^2)^{-1/2}\ .$$

Thus, if we let

$$q = \sqrt{B^2 + 4A}$$

then

$$Int = \int \pm(A + Bu - u^2)^{-1/2}\, du = \arccos\left(\frac{2u - B}{q}\right) + const\ .$$

Recalling the substitution $r = 1/u$ we obtain

$$\theta - \theta_0 = \arccos\left(\frac{1}{q}(\frac{2}{r} - B)\right)\ ,$$

thus

$$\frac{1}{r} = \frac{B}{2}\left(1 + \frac{q}{B}\cos(\theta - \theta_0)\right)\ .$$

We now define ε and a by

$$\varepsilon = \frac{q}{B}, \quad a(1 - \varepsilon^2) = \frac{2}{B}$$

and find that

$$r(\theta) = \frac{a(1 - \varepsilon^2)}{1 + \varepsilon\cos(\theta - \theta_0)}\ .$$

Using the condition $r(0) = a(1 - \varepsilon)$ (see (4.10)) we obtain $\theta_0 = 0$, thus

$$r(\theta) = \frac{a(1 - \varepsilon^2)}{1 + \varepsilon\cos\theta}\ .$$

Since this equation agrees with (4.9), the orbit of the planet is an ellipse. We have derived Kepler's first law using Newton's second law and the law of gravitational attraction.

It is interesting to relate the semi–axis a to the energy E_0. We have

$$\varepsilon^2 - 1 = \frac{4A}{B^2} = \frac{2l^2 E_0}{mk^2}$$

and

$$a(1 - \varepsilon^2) = \frac{2}{B} = \frac{l^2}{mk}\ ,$$

thus

$$a = \frac{k}{2|E_0|}\ .$$

6. Kepler's Third Law

The Period T. Recall that $mr^2\dot\theta = l$, thus

$$\frac{1}{2}r^2 d\theta = \frac{l}{2m}\,dt \; .$$

Integration over one period yields a relation between the area of the ellipse ($= \pi ab$) and the total time T needed for one passage through the elliptical orbit:

$$\pi ab = \frac{l}{2m} T \; ,$$

thus

$$T = \frac{2\pi abm}{l} \; .$$

Using that

$$b^2 = a^2(1-\varepsilon^2) = \frac{al^2}{mk}$$

one obtains

$$T^2 = 4\pi^2 a^3 \frac{m}{k} \; . \tag{4.12}$$

This equation relates the time period T of the motion to the major semi–axis a of the ellipse for motion of one body in a central field. The governing equation is (4.4).

To apply this result to the two body problem, we refer to Section 2. and recall that

$$m = \frac{m_1 m_2}{m_1 + m_2}, \qquad k = G m_1 m_2 \; .$$

Therefore, (4.12) yields

$$T^2 = \frac{4\pi^2}{G(m_1+m_2)} \cdot a^3 \; .$$

We note the proportionality of T^2 and a^3, as claimed in Kepler's third law. In application to our planetary system, m_1 is the mass of the sun and m_2 the mass of a planet. The dependency of the proportionality constant on m_2 shows that Kepler's third law does not hold exactly. However, it holds approximately since m_2/m_1 is rather small for every planet. The maximal value is $m_2/m_1 \sim 0.001$ for m_2 the mass of Jupiter.

7. Time Dependence

Newton derived Kepler's laws of planetary motion, but he did not give the explicit time dependence of the position vector $\mathbf{r}(t)$ of a planet as a function of time. Following Bessel [2], we show here how Kepler's equation (4.1) can be solved via Bessel functions and the explicit time dependence can be obtained. It should be mentioned that the solution of Kepler's equation has attracted a huge amount of attention over the centuries. The book [4] by Peter Colwell gives a survey.

For the following, we refer to Figure 4.3. The figure shows a quarter of an ellipse and a quarter of a co–centric circle whose radius is the major semi–axis of the ellipse. The sun S is at a focus of the ellipse, the point C is its center and the point P_0, the perihelion, is the point on the ellipse closest to S. The ellipse has semi–axes $a > b > 0$ and eccentricity $0 < \varepsilon < 1$. The point $P = (x, y)$ denotes the position of the planet on the ellipse, i.e.,

$$\frac{x^2}{a^2} + \frac{y^2}{b^2} = 1 \ .$$

The time since the planet passed through the perihelion P_0 is denote by t, and T denotes the total time for one revolution. In astronomy, the number $\mu = 2\pi t/T$ is then called the mean anomaly.

The point Q has the same x–coordinate as P, but lies on the *circle* of radius a, centered at C. Thus, in coordinates,

$$Q = (x, ya/b), \quad P = (x, y) \ .$$

The angle α at C between the lines CP_0 and CQ is known as the eccentric anomaly. Kepler's equation (4.1), which we will derive below, relates the mean anomaly $\mu = 2\pi t/T$ to the eccentric anomaly α. If one knows the solution $\alpha = \alpha(t)$ of Kepler's equation, then Q has coordinates

$$Q = (x, ya/b) = (a\cos\alpha, a\sin\alpha)$$

and P has coordinates

$$P = (x, y) = (a\cos\alpha, b\sin\alpha) \quad \text{with} \quad \alpha = \alpha(t) \ .$$

Therefore, to get the explicit time dependence of the motion amounts to solving Kepler's equation (4.1) for $\alpha = \alpha(t)$.

Derivation of Kepler's Equation (4.1). Denote the area SP_0P by $A(t)$. Then we have

$$A(0) = 0, \quad A(T) = \pi ab$$

and $A'(t) = const$ by Kepler's second law. Therefore,

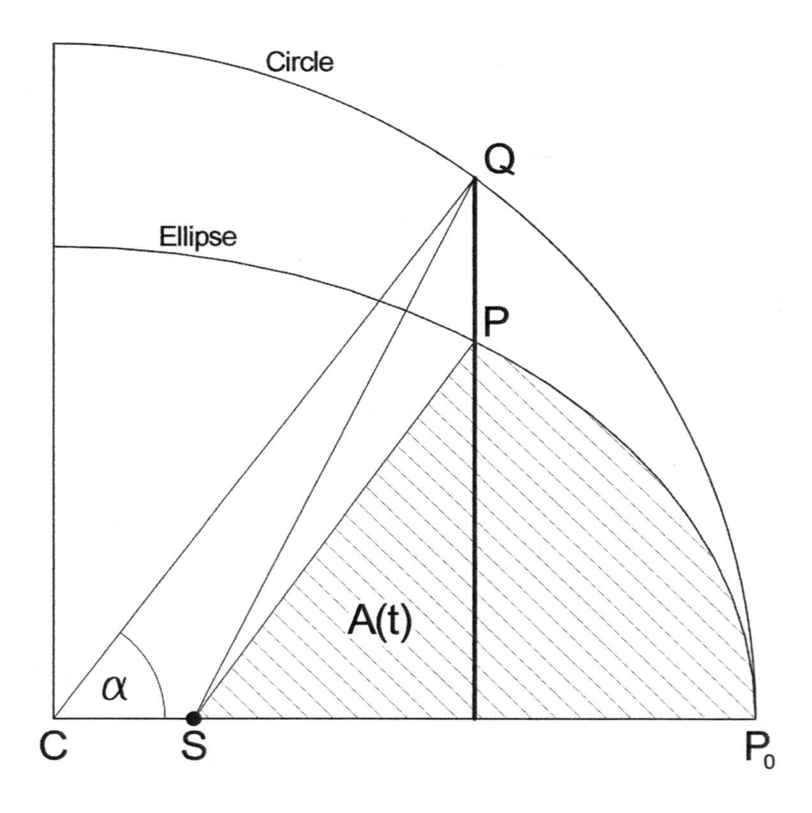

Figure 4.3. Derivation of Kepler's equation

$$A(t) = \frac{\pi abt}{T} \,.$$

Let $B(t)$ denote the area SP_0Q. Then we have

$$B(t) = \frac{a}{b}\,A(t) = \frac{\pi a^2 t}{T} \,.$$

One can also express $B(t)$ as the difference between the area of the circular sector P_0CQ and the triangle CSQ. This yields [4]

$$B(t) = \frac{1}{2}\,a^2\alpha - \frac{1}{2}\varepsilon a^2 \sin\alpha$$

and we have shown that

[4]The triangle CSQ has the base CS of length $c = \varepsilon a$ and the height $a \sin\alpha$ since the radius of the circle is a.

$$\frac{\pi a^2 t}{T} = \frac{1}{2}a^2\alpha - \frac{1}{2}\varepsilon a^2 \sin\alpha \ .$$

Dividing both sides by $a^2/2$ yields

$$\frac{2\pi t}{T} = \alpha - \varepsilon \sin\alpha \ ,$$

which is Kepler's equation.

Solution of Kepler's Equation in Terms of Bessel Functions. We follow here standard notation and denote the mean anomaly by $\mu = 2\pi t/T$ and the eccentric anomaly by u. Then Kepler's equation reads

$$\mu = u - \varepsilon \sin u \tag{4.13}$$

where $0 < \varepsilon < 1$ is a fixed constant, the eccentricity of the elliptic orbit.

Figure 4.4 shows a graph of the right–hand side of Kepler's equation, i.e., of the function $u \to u - \varepsilon \sin u$ where u varies in the interval $0 \le u \le \pi$. It is clear from the graph that, for any given $0 \le \mu \le \pi$, Kepler's equation has a unique solution $u(\mu)$ in the interval $0 \le u(\mu) \le \pi$. Using the intermediate value theorem, it is easy to give a rigorous argument for this statement because the function $f(u) = u - \varepsilon \sin u$ satisfies $f(0) = 0, f(\pi) = \pi$ and $f'(u) > 0$. Once (4.13) is solved by $u = u(\mu)$, the function $\alpha(t) = u(2\pi t/T)$ solves (4.1).

One can write down simple iterations, based on Newton's method, which quickly produce very good approximations to the exact solution $u = u(\mu)$ of (4.13). However, in Kepler's time and later, it was preferred to have an explicit formula for a solution $u = u(\mu)$ of Kepler's equation. Bessel provided such a formula, in terms of a new class of functions, now called Bessel functions .

It is clear from Figure 4.4 that $0 \le u(\mu) \le \pi$ for $0 \le \mu \le \pi$ and

$$u(0) = u(\pi) - \pi = 0 \ .$$

Therefore, we can write

$$u(\mu) - \mu = \sum_{n=1}^{\infty} c_n \sin(n\mu) \tag{4.14}$$

with Fourier coefficients c_n, which we will now determine. The result will be the somewhat incomprehensible, but explicit, formula (4.16).

Multiply (4.14) by $\sin(j\mu)$ and integrate over $0 \le \mu \le \pi$ to obtain

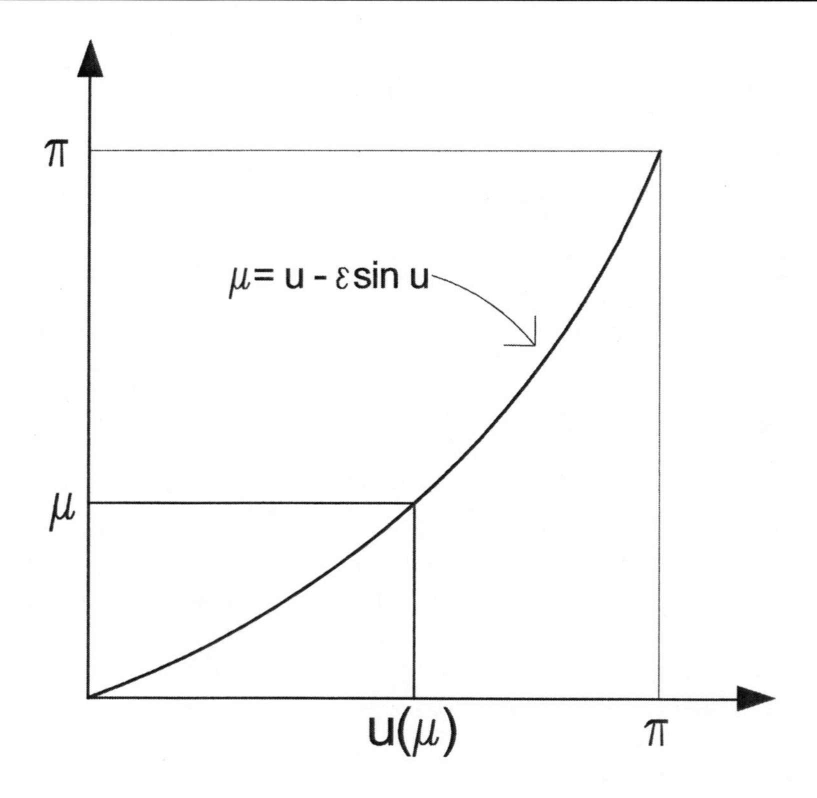

Figure 4.4. Solution of equation (4.13)

$$\begin{aligned}
\frac{\pi}{2} c_j &= \int_0^\pi (u(\mu) - \mu) \sin(j\mu)\, d\mu \\
&= \frac{1}{j} \int_0^\pi (u'(\mu) - 1) \cos(j\mu)\, d\mu \\
&= \frac{1}{j} \int_0^\pi u'(\mu) \cos(j\mu)\, d\mu
\end{aligned} \tag{4.15}$$

We now make a simple, but crucial, substitution in the last integral. First, note that, by definition, the function $\mu \to u(\mu)$ is the inverse of the function

$$v \to v - \varepsilon \sin v =: \mu(v) \ .$$

(Both functions, $u(\mu)$ and $\mu(v)$, are one–to–one and map the interval $[0, \pi]$ onto itself.) We have

$$u(\mu(v)) \equiv v, \quad u'(\mu(v))\mu'(v) \equiv 1, \quad u'(\mu)d\mu = dv \ .$$

If we now substitute $\mu = \mu(v)$ in the integral (4.15) we obtain

$$\begin{aligned} \frac{\pi}{2}c_j &= \frac{1}{j}\int_0^{\pi} \cos(j\mu(v))\, dv \\ &= \frac{1}{j}\int_0^{\pi} \cos(jv - j\varepsilon\sin(v)))\, dv \end{aligned}$$

which gives the following formula for the Fourier coefficients c_j in (4.14):

$$c_j = \frac{2}{\pi j}\int_0^{\pi} \cos\Big(j\varepsilon\sin(v) - jv\Big)\, dv \ . \tag{4.16}$$

Bessel defined the functions

$$J_j(x) = \frac{1}{\pi}\int_0^{\pi} \cos\Big(x\sin(v) - jv\Big)\, dv \quad \text{for} \quad j = 0, 1, \ldots \tag{4.17}$$

and the above expression for c_j becomes

$$c_j = \frac{2}{j}\, J_j(j\varepsilon) \ .$$

With these newly defined Bessel functions $J_j(x)$ and formula (4.14) the solution of Kepler's equation (4.13) becomes

$$u(\mu) = \mu + 2\sum_{n=1}^{\infty} \frac{1}{j}\, J_j(\varepsilon j)\sin(j\mu) \ . \tag{4.18}$$

8. Bessel Functions via a Generating Function: Integral Representation

In most of today's texts, the Bessel functions $J_j(x)$ are not defined via the integral representation (4.17). Let us derive the formula (4.17) if one defines the Bessel functions via the generating function

$$g(z,t) := \exp\Big(\frac{z}{2}(t - \frac{1}{t})\Big) = \sum_{n=-\infty}^{\infty} J_n(z)t^n \ . \tag{4.19}$$

Here z and t are complex numbers, $t \neq 0$. For fixed z, the function $t \to g(z,t)$ is analytic in $\mathbb{C} \setminus \{0\}$ and, therefore, has a unique Laurent expansion about $t = 0$. The Bessel functions $J_n(z)$ are defined as the coefficients. Note that for z and t real, the value $g(z,t)$ is real. This implies that $J_n(z)$ is real for real z.

Lemma 4.1. *Let $f(t) = \sum_{n=-\infty}^{\infty} c_n t^n$ denote a function which is analytic for $t \neq 0$ and which is real for real t. Then the Laurent coefficients c_n are all real.*

Proof: Set $g(t) = \sum \bar{c}_n t^n$ where $\bar{c}_n$ denotes the complex conjugate of c_n. Then $g(t)$ is also analytic for $t \neq 0$. Furthermore, for real t,

$$f(t) = \bar{f}(t) = g(t) \; .$$

The identity theorem for analytic functions yields that f and g are identical. Uniqueness of the Laurent coefficients c_n implies that $c_n = \bar{c}_n$ is real. ◇

In (4.19) substitute

$$t = e^{iv}, \quad t - \frac{1}{t} = 2i \sin v \; ,$$

to obtain

$$\exp(iz \sin v) = \sum J_n(z) e^{inv} \; .$$

Multiplying by e^{-ijv} and integrating over $0 \leq v \leq 2\pi$ yields

$$\int_0^{2\pi} e^{iz \sin v - ijv} \, dv = 2\pi J_j(z) \; .$$

Let $z = x$ be real and take real parts to find that

$$2\pi J_j(x) = \int_0^{2\pi} \cos(x \sin v - jv) \, dv \; .$$

If we set

$$q(v) = \cos(x \sin v - jv)$$

then

$$\begin{aligned} q(2\pi - v) &= \cos(-x \sin v + jv) \\ &= \cos(x \sin v - jv) \\ &= q(v) \end{aligned}$$

which yields that

$$\pi J_j(x) = \int_0^{\pi} \cos(x \sin v - jv) \, dv \; .$$

This result agrees with Bessel's original definition (4.17).

9. Discussion

The mathematics in this chapter should make it clear what one means by an explicit solution of a non–trivial initial value problem, the two body problem. In principle, using the formula (4.18) and other formulas of this chapter, one obtains the position of a planet as a function of time as a convergent series. Of course, it was tried to go from two bodies to three and more. One can write down a system of equations, extending (4.2) and (4.3), which determines the motion of n bodies. However, it turns out that one can no longer find solutions in terms of convergent series if $n \geq 3$. The interesting history of this subject — one of the starting points of chaos theory — is described in [5], for example. For old (due to Leonard Euler (1707–1783) and Joseph–Louis Lagrange (1736–1813)) and new solutions of the three–body problem, see [14].

Is it possible, in principle, to describe the motion of everything in the universe by a system $u' = f(u)$? If this would be so, then the universe would run like an exact clock. We feel that we have free will, and how we decide does influence the future. If the model $u' = f(u)$ would be valid for everything, then our free will would just be an illusion.

It is now easy to dismiss the view going with a deterministic system $u' = f(u)$. In fact, quantum physics teaches us that there are random events. Even in principle, their occurrence cannot be predicted exactly. We can only calculate their probabilities. Here the possibility of free will enters the door. At least, its existence does not *contradict* physical law. Nevertheless, how the free, which we feel, interacts with matter remains unclear. Can our free will influence quantum events? Or do quantum events form our free will?

The eminent physicists Stephen Hawking [5] and Erwin Schrödinger [6] do not agree on free will. In their popular book [8] Hawking and Mlodinow write *free will is just an illusion.* Their main argument, it seems, is that quantum mechanics determines the probabilities of micro events, and there is no physical basis for influencing which of the possible events are realized. In contrast, in his essay [15] Erwin Schrödinger writes that there is *incontrovertible direct experience* that humans have free will.

The non–determinism offered by quantum mechanics seems to be necessary, but not sufficient for the existence of free will. It also seems to be necessary that there is a non–physical entity (the *I* or the *mind*) which can act on physical reality. Does such an entity exist? It is not obvious that our mind will ever be able to understand how our own mind works.

There is a huge amount of literature in theology and philosophy about free will. The basic problem is this: Unless we have free will, how can we be responsible for our actions? On the other hand, if God knows everything, from beginning to end, how can there be any freedom on our part? The author is not at all an expert on these issues, but his free will

[5] British physicist and cosmologist born in 1942

[6] Austrian physicist (1878–1961), Nobel Prize in Physics in 1933. Erwin Schrödinger was one of the founders of quantum mechanics.

still allows him to state his opinion: God, of course, *could* determine all our actions, but *chose* not to do so. He gave us free will since, otherwise, the universe would be too boring.

Chapter 5

Is Time Reversible?

Summary: The equations for the two–body problem are used to illustrate time reversibility. We then give a general definition of this concept. What we see around us is not time reversible at all. Is this related to the special initial condition in which the universe started?

1. Reversibility for the Two Body Problem

What happens as time progresses is typically irreversible. You can jump into a deep swimming pool, but can you jump back?

There are important exceptions to irreversibility. Consider, for example, the two body problem, which we discussed in Chapter 4. The governing equations for the two position vectors $\mathbf{r}_1(t)$ and $\mathbf{r}_2(t)$ are

$$m_1\ddot{\mathbf{r}}_1 = Gm_1m_2\frac{\mathbf{r}_2 - \mathbf{r}_1}{|\mathbf{r}_2 - \mathbf{r}_1|^3} \tag{5.1}$$

$$m_2\ddot{\mathbf{r}}_2 = Gm_1m_2\frac{\mathbf{r}_1 - \mathbf{r}_2}{|\mathbf{r}_1 - \mathbf{r}_2|^3} \tag{5.2}$$

and if we give initial conditions

$$\mathbf{r}_1(0) = \mathbf{r}_{10}, \quad \dot{\mathbf{r}}_1(0) = \dot{\mathbf{r}}_{10}$$
$$\mathbf{r}_2(0) = \mathbf{r}_{20}, \quad \dot{\mathbf{r}}_2(0) = \dot{\mathbf{r}}_{20}$$

then a unique solution $\mathbf{r}_1(t), \mathbf{r}_2(t)$ is determined for all time.[1] Here $\mathbf{r}_{10}, \mathbf{r}_{20}, \dot{\mathbf{r}}_{10}, \dot{\mathbf{r}}_{20} \in \mathbb{R}^3$ are any fixed vectors.

[1] Let us ignore the exceptional cases where the two bodies collide.

Now suppose we change the signs of the initial velocities and give the initial condition

$$\begin{aligned} \mathbf{s}_1(0) = \mathbf{r}_{10}, \quad \dot{\mathbf{s}}_1(0) = -\dot{\mathbf{r}}_{10} \ , \\ \mathbf{s}_2(0) = \mathbf{r}_{20}, \quad \dot{\mathbf{s}}_2(0) = -\dot{\mathbf{r}}_{20} \ . \end{aligned}$$

We write $\mathbf{s}_1(t)$ and $\mathbf{s}_2(t)$ instead of $\mathbf{r}_1(t)$ and $\mathbf{r}_2(t)$ to make clear that new functions must be determined. The differential equations for $\mathbf{s}_1(t)$ and $\mathbf{s}_2(t)$ are the same as those for $\mathbf{r}_1(t)$ and $\mathbf{r}_2 9t)$, however. Only the initial condition is changed. It is easy to check that the system for $\mathbf{s}_1(t)$ and $\mathbf{s}_2(t)$ is solved by

$$\mathbf{s}_1(t) = \mathbf{r}_1(-t), \quad \mathbf{s}_2(t) = \mathbf{r}_2(-t) \ .$$

We see that for every solution $\Big(\mathbf{r}_1(t), \mathbf{r}_2(t)\Big)$ of the two body problem, there is another solution, namely $\Big(\mathbf{r}_1(-t), \mathbf{r}_2(-t)\Big)$, which traces out the same orbit, but goes backwards through the orbit as time progresses. For the two body problem, time has no preferred direction. The problem is time reversible.

2. Reversibility: General Definition

Below we give a formal definition of time reversibility of a first order system $u' = f(u)$. Our definition is motivated by the following simple result.

Lemma 5.1. *Consider a system $u' = f(u)$ where $f : \mathbb{R}^n \to \mathbb{R}^n$ is a smooth map. Furthermore, let $R : \mathbb{R}^n \to \mathbb{R}^n$ denote a linear map satisfying $R^2 = I =$ identity. If*

$$-Rf(w) = f(Rw) \quad \textit{for all} \quad w \in \mathbb{R}^n \tag{5.3}$$

and if the vector function $u(t)$ satisfies $u' = f(u)$, then the function $v(t) = Ru(-t)$ satisfies $v' = f(v)$.

Proof: This follows from

$$\begin{aligned} v'(t) &= -Ru'(-t) \\ &= -Rf(u(-t)) \\ &= f(R(u(-t)) \\ &= f(v(t)) \end{aligned}$$

◇

Definition 5.1: Using the notations of the previous lemma, we call the system $u' = f(u)$ time reversible w.r.t. the operator R if condition (5.3) holds, i.e., if every solution $u(t)$ produces another solution $v(t) = Ru(-t)$ of the system.

To apply the definition to the second order system (5.1), (5.2), we write it in first order form using the variables $\mathbf{r}_1, \dot{\mathbf{r}}_1, \mathbf{r}_2, \dot{\mathbf{r}}_2$. The operator R then multiplies the velocity vectors by -1.

Time reversibility holds in the same way for n bodies, instead of two, as long as gravitational attraction is the only law of interaction. In fact, time reversibility holds more generally in micro physics. This goes under the name CPT symmetry: The dynamical equations of the standard model are symmetric under time reversal if one also applies parity and charge conjugation. See, for example, [6].

3. Discussion

We may speculate that the evolution of the whole universe can be described by a system $u' = f(u)$, which is time reversible w.r.t. an appropriate operator R. (This is an extreme form of determinism.) A natural question, then, is how irreversibility can arise, which we observe in the macro world. If the general law, $u' = f(u)$, is time reversible, the observed irreversibility may come from the very special initial state in which the universe started, expressed by the big bang theory. In this view, the fact that you can jump into the swimming pool is ultimately still a consequence of the big bang. Had the universe started in thermodynamic equilibrium, nothing of much interest could ever happen.

Our view of the past and the future also strongly relies on an irreversible evolution of the universe because of a special initial state. If we find a dinosaur bone, we expect it to crumble to dust in the distant future, but we believe that it belonged to a live dinosaur in the distant past. Before the big bang theory was developed, Ludwig Boltzmann (1844–1906) expressed the view that what we observe is only due to an *accidental* fluctuation away from thermodynamic equilibrium. If that would be correct, we should believe that the dinosaur bone was formed accidentally from dust in the past, similarly as it will crumble to dust in the future, just time reversed. A fluctuation back to a live dinosaur would be much, much less likely.

Irreversibility in time (of macro physics) is closely related to the 2nd law of thermodynamics. Originally, this law dealt with conversion between heat energy and macroscopic mechanical energy. Mechanical energy gets converted to heat through friction without any problem. The opposite transformation is more difficult and always has limited efficiency. A first and very interesting insight into these issues was obtained by Sadi Carnot (1796–1832). His work was later formalized, in various forms, as the 2nd law of thermodynamics. We will present Carnot's ideas in Section 4..

Chapter 6

The Bernoulli Shift and the Logistic Map

Summary: This chapter has two aims: First, to give an example of a dynamical system with *sensitive dependence on initial conditions*, the Bernoulli shift. The corresponding map $y \to B(y)$ is quite simple (see Figure 6.1), but the discrete–time dynamics $y_n \to B(y_n) = y_{n+1}$ has many interesting features. The Lyapunov exponent of any trajectory is positive, implying sensitive dependence on initial conditions.

For any dynamical system with sensitive dependence on initial conditions, the long–time prediction of individual trajectories is practically limited since an error in the initial condition typically grows exponentially in time. Nevertheless, and to show this is our second aim, the *average* behavior of the system may still be very well–determined. We use the dynamics of the logistic map $x \to 4x(1-x)$ and its relationship to the Bernoulli shift to illustrate this point.

1. The Bernoulli Shift: Definition

Define the map $B : [0,1) \to [0,1)$ by

$$B(y) = \begin{cases} 2y & \text{for} \quad 0 \le y < \frac{1}{2}\ , \\ 2y-1 & \text{for} \quad \frac{1}{2} \le y < 1\ . \end{cases}$$

We can also write $B(y) = (2y) \bmod 1$.

The discrete–time dynamics $y_n \to B(y_n)$ is called the Bernoulli shift.

If we choose a starting value $y_0 \in [0,1)$ then the iteration

$$y_{n+1} = B(y_n), \quad n = 0, 1, 2, \ldots$$

determines a trajectory in the state space $[0,1)$, which we denote by $\{y_n\}_{n=0,1,\ldots}$ or simply by y_n.

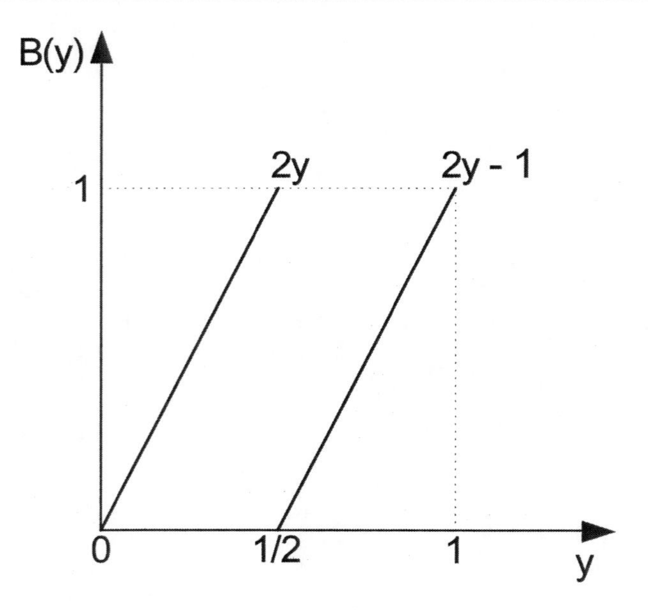

Figure 6.1. Graph of the function $B(y)$

To study the Bernoulli shift, the *binary representation* of numbers y is very useful.

Lemma 6.1. *Every $0 \le y < 1$ can be written in the form*

$$y = \sum_{j=1}^{\infty} b_j \, 2^{-j} \quad \textit{where} \quad b_j = 0 \quad \textit{or} \quad b_j = 1 \; . \tag{6.1}$$

This representation of y is unique if one forbids that $b_j = 1$ for all large j. Conversely, any series of the form [1] *(6.1) determines a number $y \in [0, 1)$.*

One can prove this result by modifying the arguments which lead to the familiar decimal representation

$$y = \sum_{j=1}^{\infty} d_j \, 10^{-j} \quad \text{where} \quad d_j \in \{0, 1, 2, \ldots, 9\} \; .$$

If (6.1) holds then

$$y = [0.b_1 b_2 b_3 \ldots]_{base2}$$

is called the binary representation of y. Application of B yields

[1]The exceptional case where $b_j = 1$ for all large j will always be excluded.

$$B(y) = [0.b_2 b_3 b_4 \ldots]_{base2} \; .$$

Thus, application of B just shifts the binary point one place to the right and replaces b_1 by 0.

Example: Consider the number

$$y_0 = [0.001\,001\,001\,\ldots]_{base2} \; .$$

We have

$$y_1 = [0.010\,010\,010\,\ldots]_{base2}$$

and

$$y_2 = [0.100\,100\,100\,\ldots]_{base2}$$

and $y_3 = y_0$. Thus, the numbers y_0, y_1, y_2 form an orbit of period 3 of B; we have a so–called 3–cycle.

In standard notation,

$$y_0 = \frac{1}{8} \sum_{j=0}^{\infty} \frac{1}{8^j} = \frac{1}{8} \cdot \frac{1}{1 - \frac{1}{8}} = \frac{1}{7} \; .$$

Application of $B(y) = (2y) \bmod 1$ yields

$$y_0 = \frac{1}{7}, \quad y_1 = \frac{2}{7}, \quad y_2 = \frac{4}{7}, \quad y_3 = \frac{8}{7} - 1 = \frac{1}{7} = y_1 \; .$$

This confirms that y_0, y_1, y_2 form a 3–cycle.

2. The Bernoulli Shift: Dynamical Properties

Lyapunov Exponent: Let us determine the Lyapunov exponent for the trajectories of B. See Definition 3.1 in Section 3.. If y_n is a trajectory and $y_n \neq \frac{1}{2}$ for all n then formula (3.3) yields $\lambda = \log 2$ since $B'(y_n) = 2$ for all n.

If y_n is a trajectory and $y_n = \frac{1}{2}$ for some $n = n_0$, then Definition 3.1 cannot be applied directly since the map $B(y)$ is not differentiable at $y = \frac{1}{2}$. However, if $y_{n_0} = \frac{1}{2}$, then $y_n = 0$ for all $n > n_0$. If one replaces the derivative $\Phi'(u_j)$, which occurs in Definition 3.1, by a one–sided derivative, one still finds that $\lambda = \log 2$ since $B'(0) = 2$.

Let us recall from Section 3. what it means that $\lambda = \log 2$: In one time step, a small error in y_0 will grow approximately by the factor

$$e^{\lambda} = e^{\log 2} = 2 \; .$$

Let us confirm this directly for the Bernoulli shift: Consider two initial values

$$
\begin{aligned}
y_0 &= [0.b_1 \ldots b_N b_{N+1} b_{N+2} \ldots]_{base2} \\
y_0 + \delta &= [0.b_1 \ldots b_N c_{N+1} c_{N+2} \ldots]_{base2}
\end{aligned}
$$

with $b_{N+1} \neq c_{N+1}$. The initial error is

$$\delta = \sum_{j=N+1}^{\infty} (c_j - b_j) 2^{-j} \ .$$

An application of B shifts every digit to the left. Therefore, after one time step, the error δ is multiplied by 2 (exactly). It is not difficult to show the following:

Lemma 6.2. *Given any $0 \le y_0 < 1$, any accuracy $\varepsilon > 0$, and any $N \ge 1$, there is a perturbation $y_0 + \delta$ of y_0 so that*

$$0 < |\delta| \le \varepsilon \quad \textit{and} \quad |B^n(y_0) - B^n(y_0 + \delta)| = 2^n |\delta| \quad \textit{for} \quad n = 1, 2, \ldots N \ .$$

This is a precise statement of error doubling that occurs; it implies sensitive dependence on the initial condition.

Three Sets of Initial Values: Consider the following three subsets S_j of the state space $[0,1)$:

$$
\begin{aligned}
S_1 &= \{y \in [0,1) \ : \ y = \sum_{j=1}^{J} b_j 2^{-j}, \ J \text{ finite}\} \\
S_2 &= \{y \in [0,1) \ : \ y = \sum_{j=1}^{\infty} b_j 2^{-j}, \ b_j \text{ is periodic for large } j\} \\
S_3 &= [0,1) \setminus S_2
\end{aligned}
$$

The set S_2 is the set of all rationals in the interval $[0,1)$. Both sets S_1 and S_2 are denumerable, i.e., countably infinite. All three sets S_j are dense in $[0,1)$.

Now consider a trajectory $y_n = B^n(y_0)$ and its long time behavior. If $y_0 \in S_1$ then $y_n = 0$ for all large n. The point $y = 0$ is the only fixed point of B. If $y_0 \in S_2 \setminus S_1$ then y_n becomes a periodic cycle for large n . If $y_0 \in S_3$ then y_n does neither approach a fixed point nor a periodic cycle. Since all three sets are dense in the interval $[0,1)$ we see that the long time behavior of a trajectory depends, in some sense, *discontinuously* on the initial value y_0. Infinite accuracy in y_0 is required if one wants to know the long time fate of the trajectory y_n.

We also note that, given any $q \in \{2, 3, \ldots\}$, the Bernoulli shift has a cycle of length q . Every cycle, as well as the unique fixed point $y = 0$, are unstable: There are arbitrarily small perturbations leading away from the cycle or the fixed point.

Here is an interesting question: What happens if you implement the Bernoulli shift on a computer? Try it and explain!

3. The Logistic Map and Its Relation to the Bernoulli Shift

If $0 \leq r \leq 1$ is a parameter, then the map

$$x \to rx(1-x), \quad 0 \leq x \leq 1 , \tag{6.2}$$

maps the unit interval $I = [0,1]$ into itself. Here we consider only the case $r = 4$ and then call (6.2) the logistics map. (To study the r–dependence of the dynamics of (6.2) is an interesting subject of bifurcation theory. The famous Feigenbaum [2] sequence of period doubling occurs.)

The following lemma relates the logistic map to the Bernoulli shift via the coordinate transformation

$$x = \sin^2(\pi y) =: s(y), \quad 0 \leq y \leq 1 . \tag{6.3}$$

Lemma 6.3. *Let $f(x) = 4x(1-x)$ and $B(y) = (2y) \, mod 1$. Let $y_0 \in [0,1)$ and set $y_1 = B(y_0)$. Also, let $x_0 \in [0,1]$ and set $x_1 = f(x_0)$. With these settings we claim that $x_0 = s(y_0)$ implies $x_1 = s(y_1)$.*

Proof: We have

$$\begin{aligned} x_1 &= 4x_0(1-x_0) \\ &= 4s(y_0)(1-s(y_0)) \\ &= 4\sin^2(\pi y_0)\cos^2(\pi y_0) \\ &= \sin^2(2\pi y_0) \\ &= \sin^2(\pi((2y_0) \bmod 1)) \\ &= \sin^2(\pi y_1) \\ &= s(y_1) \end{aligned}$$

◇

An application of this lemma and our results on the Bernoulli shift yields the following for the logistic map:

a) There is a dense denumerable set $T_1 \subset [0,1]$ so that $x_0 \in T_1$ implies $x_n \to 0$.

[2]named after the American mathematical physicist Mitchel Feigenbaum (1944 –)

b) There is a dense denumerable set $T_2 \subset [0,1]$ so that $x_0 \in T_2$ implies that x_n becomes a periodic cycle for large n.

c) There are periodic cycles of any length.

d) There is a dense set T_3 of full (Lebesgue) measure so that x_n does not become periodic if $x_0 \in T_3$.

e) The dynamics of the logistic map has sensitive dependence on the initial condition.

4. Average Behavior of the Logistic Map

The above results make it clear that the precise fate of any trajectory x_n of the logistic map

$$x \to f(x) = 4x(1-x)$$

cannot be determined unless one works in exact arithmetic. It is instructive to see what happens if one implements the dynamics on a computer.

For example, if you choose $x_0 = 100/777$, compute the first 10,000 iterates x_n, and plot their histogram based on 200 subintervals of the unit interval, you will get Figure 6.2.

Here is the simple Matlab code:

```
% Histogram for logistic map
N=10000;
r=4.;
x(1)=100/777;
for j=1:N
x(j+1)=r*x(j)*(1-x(j));
end
hist(x,200)
title('Histogram of f(x)=4x(1-x)')
```

If you choose another starting value x_0 or use Matlab's random number generator and set $x_0 = rand$, it is very likely that you will obtain almost the same histogram. This indicates that the *average behavior* of the trajectories does not depend on the chosen initial value. In this section, we try to explain this.

In fact, we will show that the dynamics of the logistic map has an invariant measure with density [3]

$$H(x) = \frac{1}{\pi} \cdot \frac{1}{\sqrt{x(1-x)}}, \quad 0 < x < 1 \, . \tag{6.4}$$

[3]The measure with density $H(x)$ is called invariant under the map f if $\int_I H(x)dx = \int_{f^{-1}(I)} H(x)dx$ for all intervals $I \subset [0,1]$. Here, by definition, the set $f^{-1}(I)$ consists of all $\xi \in [0,1]$ with $f(\xi) \in I$.

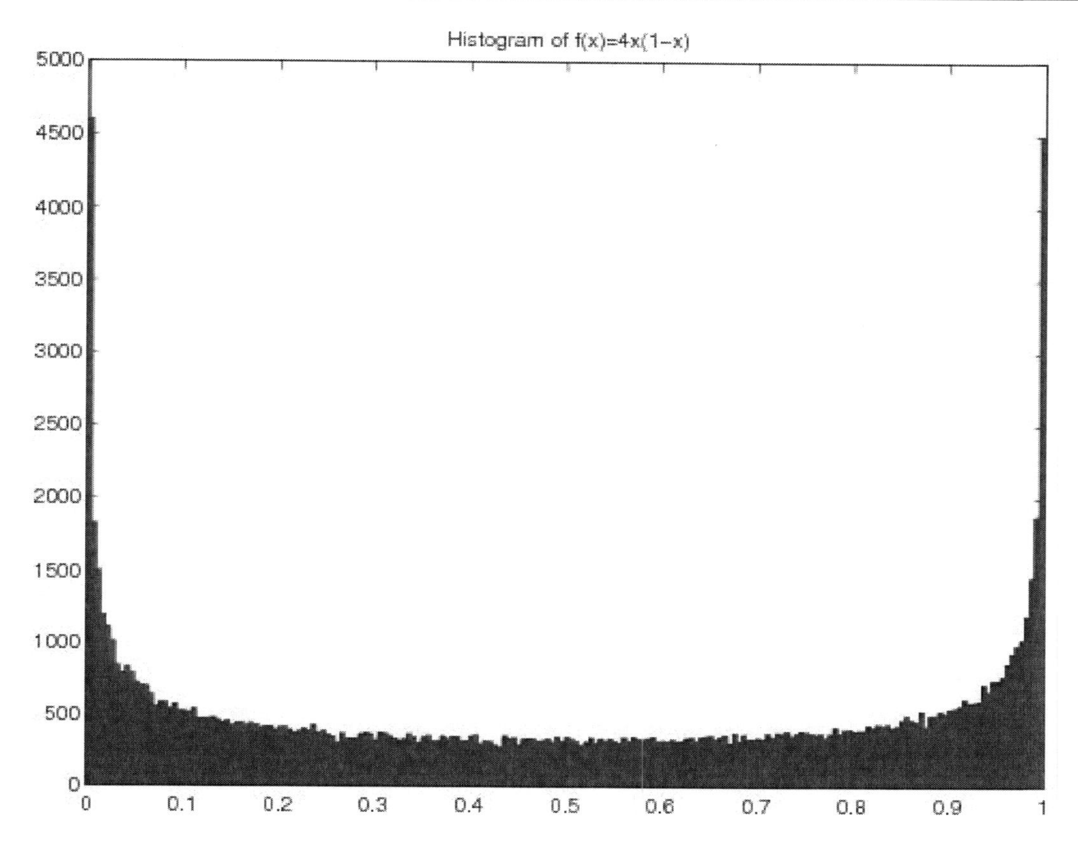

Figure 6.2. Histogram for the Logistic Map

The function $H(x)$, properly scaled, gives an approximation to the shape of the above histogram.

Derivation of $H(x)$: Assume that, if $x_0 \in [0,1]$ is randomly[4] chosen, then the orbit $x_n = f^n(x_0)$ has the following property: We can express the probability that $x_n \in [a,b]$ as an integral:

$$prob(x_n \in [a,b]) = \int_a^b H(p)\,dp\ . \tag{6.5}$$

Here the function $H : (0,1) \to [0,\infty)$ is assumed to be smooth, independent of the interval $[a,b]$, and independent of the index n. Also, since $f(x) \equiv f(1-x)$, we make the reasonable symmetry assumption $H(x) = H(1-x)$. Then, using (6.5),

[4]It is not trivial to make mathematically precise what this means. For practical purposes, let's assume that $x_0 = rand$ using Matlab's random number generator.

$$prob(x_n \in [0,x] \cup [1-x,1]) = 2 \int_0^x H(p)\, dp \quad \text{for} \quad 0 < x \le \frac{1}{2} \, .$$

The points $x_n \in [0,x] \cup [1-x,1]$ are precisely those with

$$x_{n+1} \in [0, f(x)] \, ,$$

therefore,

$$prob(x_n \in [0,x] \cup [1-x,1]) = prob(x_{n+1} \in [0, f(x)]) \, .$$

(See Figure 6.3.) Using the assumption that H does not depend on n we obtain that

$$2 \int_0^x H(p)\, dp = \int_0^{f(x)} H(p)\, dp \, .$$

Differentiations yields the functional equation

$$2H(x) = f'(x) H(f(x)) \, ,$$

thus

$$H(x) = (2-4x) H(4x(1-x)) \, . \tag{6.6}$$

Also, H must satisfy the normalization condition

$$\int_0^1 H(x)\, dx = 1 \, .$$

Lemma 6.4. *The function*

$$H(x) = \frac{1}{\pi} \cdot \frac{1}{\sqrt{x(1-x)}}$$

solves the functional equation and the normalization condition.

Proof: We first ignore the normalization condition and set

$$h(y) = \frac{1}{\sqrt{y(1-y)}} \, .$$

Then we have

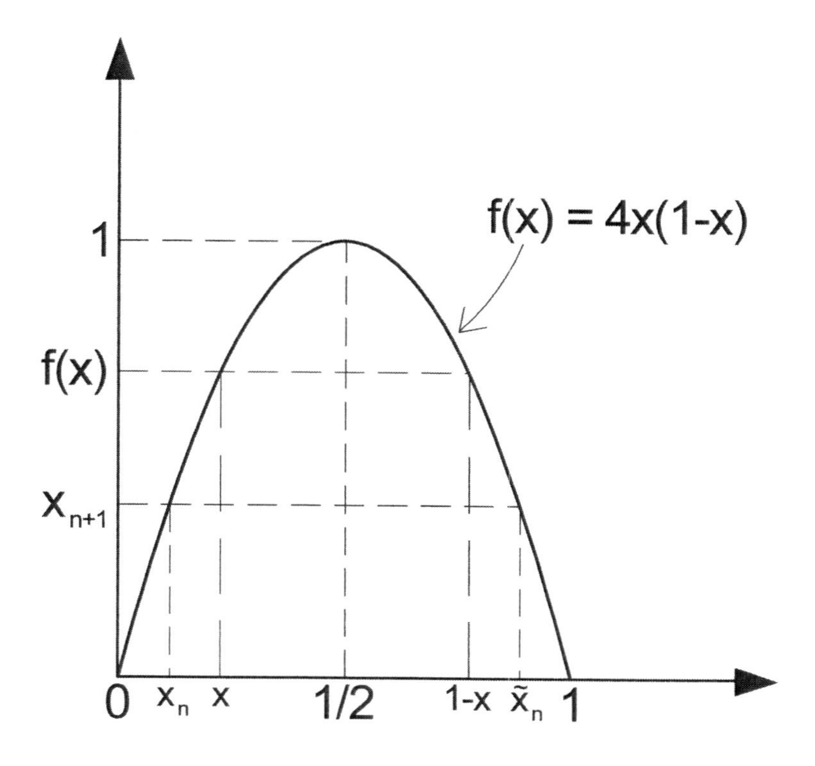

Figure 6.3. $x_n \in [0, x], \tilde{x}_n \in [1-x, 1]$ and $x_{n+1} \in [0, f(x)]$

$$
\begin{aligned}
(2-4x)h(4x(1-x)) &= \frac{2-4x}{\sqrt{4x(1-x)(1-4x(1-x))}} \\
&= \frac{1-2x}{\sqrt{x}}\Big((1-x)(1-4x+4x^2)\Big)^{-1/2} \\
&= \frac{1-2x}{\sqrt{x}}\Big((1-x)(1-2x)^2)\Big)^{-1/2} \\
&= \frac{1}{\sqrt{x(1-x)}} \\
&= h(x)
\end{aligned}
$$

This shows that $h(x)$ solves the functional equation.

Lemma 6.5.

$$\int_0^1 \frac{dx}{\sqrt{x(1-x)}} = \pi$$

Proof: Set

$$g(x) = \arctan\sqrt{\frac{1}{x}-1} \quad \text{for} \quad 0<x<1$$

and recall that $\arctan'(y) = (1+y^2)^{-1}$. We compute

$$\begin{aligned} g'(x) &= \frac{1}{1+\frac{1}{x}-1}\,\frac{1}{2}\Big(\frac{1}{x}-1\Big)^{-1/2}(-x^{-2}) \\ &= -\frac{1}{2x}\,\frac{1}{\sqrt{\frac{1}{x}-1}} \\ &= -\frac{1}{2}\,\frac{1}{\sqrt{x(1-x)}} \end{aligned}$$

Therefore, the function $-2g(x)$ has the derivative $h(x)$ for $0<x<1$.

For $\varepsilon > 0$:

$$\begin{aligned} \int_\varepsilon^1 \frac{dx}{\sqrt{x(1-x)}} &= -2(g(1)-g(\varepsilon)) \\ &= 2g(\varepsilon) \\ &= 2\arctan\sqrt{\frac{1}{\varepsilon}-1} \end{aligned}$$

As $\varepsilon \to 0$, the integral converges to π. This completes the proof of Lemma 6.5, and Lemma 6.4 also follows. $\diamond$

Remarks: 1. The assumptions that we made to derive the functional equation (6.6) are not easily justified. In fact, if one considers the dynamics $x_n \to x_{n+1} = rx_n(1-x_n)$ for $0<r<4$, then generally an invariant measure with smooth density does not exist.

2. The main point of this chapter was to show that an evolution with sensitive dependence on initial conditions may, on *average*, still behave completely predictably. The logistic map $x \to 4x(1-x)$ illustrates this. The reader may ask why the Bernoulli shift was not used as an example. If you have answered the question at the end of Section 2., this will be clear: Because of computer arithmetic, the numerical realization of the Bernoulli shift gives misleading answers.

Chapter 7

Evolution on Two Time–Scales

Summary: In this short chapter we will consider a simple example of an initial value problem whose solution varies on two time scales. The problem contains a parameter $0 < \varepsilon << 1$, and $1/\varepsilon$ is the separation of the two scales. We also comment briefly on a difficult example, the motion of two planets around the sun.

1. Fast and Slow Time Scales

First recall that the initial value problem

$$u' = iu, \quad u(0) = 1 \ ,$$

(where $i^2 = -1$) is solved by

$$u(t) = e^{it} = \cos t + i \sin t$$

and, similarly,

$$v' = \frac{i}{\varepsilon} v, \quad v(0) = 1 \ ,$$

is solved by

$$v(t) = e^{it/\varepsilon} = \cos(t/\varepsilon) + i \sin(t/\varepsilon) \ .$$

In Figure 7.1 we plot the real parts of $u(t)$ and $v(t)$ for $0 \le t \le 2\pi$ and $\varepsilon = 1/20$.

In the interval $0 \le t \le 2\pi$ the function $u(t)$ goes through one oscillation whereas $v(t)$ varies 20 times for $\varepsilon = \frac{1}{20}$. One says, somewhat vaguely, that $u(t) = e^{it}$ varies on the slow time scale whereas $v(t) = e^{it/\varepsilon}$ varies on the fast time scale. Somewhat more formally, we can introduce the variable of fast time,

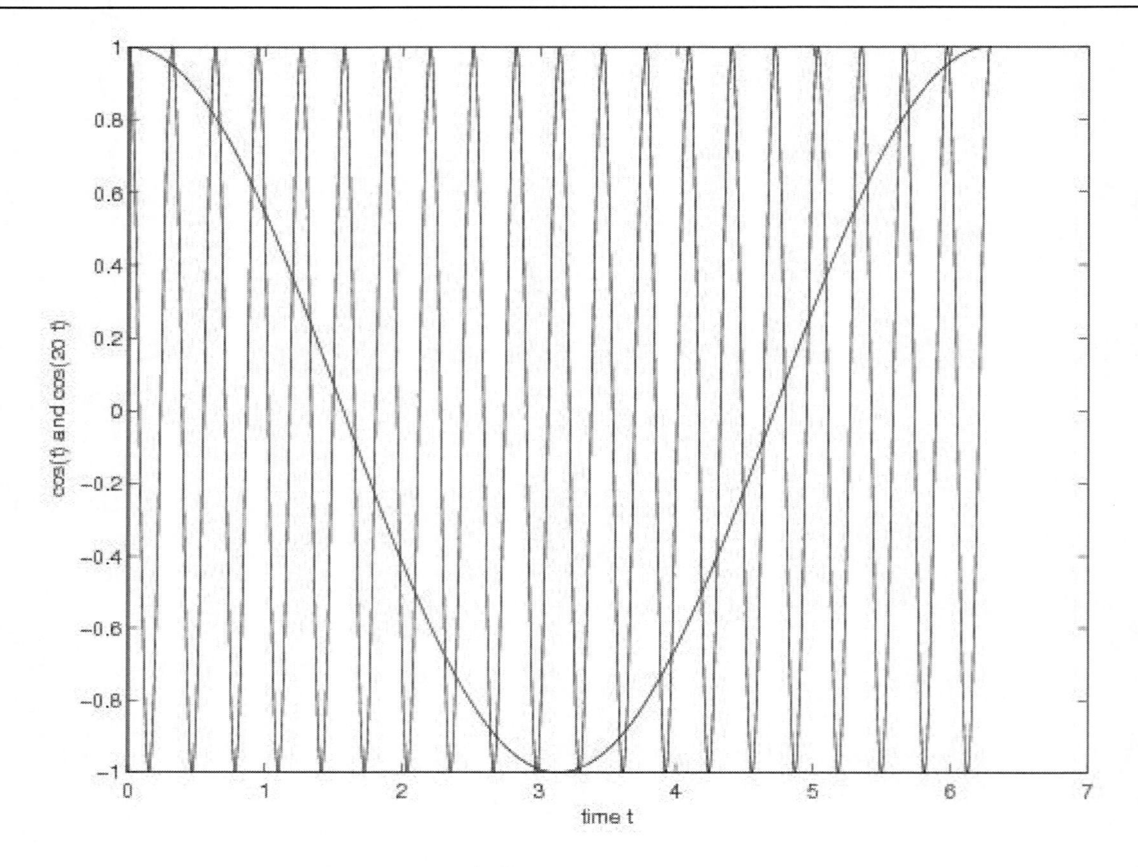

Figure 7.1. A Slow and a Fast Function

$$\tau = \frac{t}{\varepsilon} \ .$$

Then we have

$$v(t) = e^{it/\varepsilon} = e^{i\tau} \ .$$

The τ–derivative of this function is $\mathcal{O}(1)$ whereas the t–derivative is $\mathcal{O}(1/\varepsilon)$. If we only want to work with functions whose derivatives are $\mathcal{O}(1)$, we must consider v as a function of the fast time variable, τ.

2. Simple Example

Let us first recall some results for linear ODEs

$$u' = au + g(t), \quad u(0) = u_0 , \tag{7.1}$$

where $a \in \mathbb{C}$ is a constant and $g(t)$ is a continuous function. The corresponding homogeneous equation $u' = au$ has the general solution

$$u_{hom}(t) = ce^{at}$$

where c is a free parameter. The solution of the inhomogeneous equation (7.1) for $u_0 = 0$ is

$$u_{in}(t) = \int_0^t e^{a(t-s)} g(s)\, ds$$

and the solution of (7.1) for general u_0 is

$$u(t) = u_0 e^{at} + \int_0^t e^{a(t-s)} g(s)\, ds . \tag{7.2}$$

Consider the following initial value problem

$$u' = \frac{i}{\varepsilon}(u - \sin t), \quad u(0) = 1 , \tag{7.3}$$

where $0 < \varepsilon << 1$. Applying (7.2) one can write down the exact solution in closed form,

$$u(t, \varepsilon) = e^{it/\varepsilon} - \frac{i}{\varepsilon} \int_0^t e^{i(t-s)/\varepsilon} \sin s\, ds . \tag{7.4}$$

If one now uses the identity $\sin s = \frac{1}{2i}(e^{is} - e^{-is})$, an explicit solution formula can be obtained. Nevertheless, it is instructive to obtain an approximation to the solution by a formal process. The process has two steps:

Step 1: Multiply the differential equation (7.3) by ε and then set $\varepsilon = 0$ to obtain the approximation

$$U_0(t) = \sin t$$

for the solution of (7.3).

Step 2: The function $U_0(t) = \sin t$ violates the initial condition $u(0) = 1$. With an unknown function $h(t)$ write the solution $u(t)$ of (7.3) as

$$u(t) = \sin t + h(t)$$

and obtain that

$$\cos t + h'(t) = \frac{i}{\varepsilon} h(t), \quad h(0) = 1 .$$

We can also write this as

$$h'(t) = \frac{i}{\varepsilon}\,(h(t) - \frac{\varepsilon}{i}\,\cos t), \quad h(0) = 1 \ .$$

Now we neglect the term $-(\varepsilon/i)\cos t$ and obtain the approximation

$$h_0(t) = e^{it/\varepsilon}$$

for $h(t)$. Together, the two steps yield the function

$$u_0(t,\varepsilon) = \sin t + e^{it/\varepsilon}$$

as an approximation for the exact solution $u(t,\varepsilon)$ of (7.3).

Note that $u_0(t,\varepsilon)$ is a sum of a function varying on the slow time scale t and a function varying on the fast time scale $\tau = t/\varepsilon$.

We will now prove that the function $u_0(t,\varepsilon) = \sin t + e^{it/\varepsilon}$ approximates the exact solution $u(t,\varepsilon)$ to order $\mathcal{O}(\varepsilon)$, uniformly for all time:

Theorem 7.1. *Let $u(t,\varepsilon)$ denote the exact solution of (7.3). Then we have*

$$|u(t,\varepsilon) - (\sin t + e^{it/\varepsilon})| \le 4\varepsilon \quad \textit{for all} \quad t \in \mathbb{R}$$

if $0 < \varepsilon \le \frac{1}{2}$.

Proof: Denote the error by

$$\delta(t) = \delta(t,\varepsilon) = u(t,\varepsilon) - \sin t - e^{it/\varepsilon} \ .$$

We have

$$\begin{aligned}
\delta'(t) &= u'(t) - \cos t - \frac{i}{\varepsilon}\,e^{it/\varepsilon} \\
&= \frac{i}{\varepsilon}\,(u(t) - \sin t) - \cos t - \frac{i}{\varepsilon}\,e^{it/\varepsilon} \\
&= \frac{i}{\varepsilon}(\delta(t) + e^{it/\varepsilon}) - \cos t - \frac{i}{\varepsilon}\,e^{it/\varepsilon}
\end{aligned}$$

This yields the error equation

$$\delta' = \frac{i}{\varepsilon}\,\delta - \cos t, \quad \delta(0) = 0 \ ,$$

thus

$$\delta(t) = -e^{it/\varepsilon}\int_0^t e^{-is/\varepsilon}\cos s\,ds \ .$$

Through integration by parts,

$$\int_0^t e^{-is/\varepsilon} \cos s \, ds = -\frac{\varepsilon}{i} e^{-is/\varepsilon} \cos s \Big|_{s=0}^{s=t} - \frac{\varepsilon}{i} \int_0^t e^{-is/\varepsilon} \sin s \, ds \ . \tag{7.5}$$

We now recall that

$$|e^{i\alpha}| = 1 \quad \text{for all real } \alpha \ , \tag{7.6}$$

which yields the bound

$$\Big| -\frac{\varepsilon}{i} e^{-is/\varepsilon} \cos s|_{s=0}^{s=t} \Big| \le 2\varepsilon \ .$$

It remains to estimate the integral on the right–hand side of (7.5). Using the identity

$$\sin s = \frac{1}{2i} \left(e^{is} - e^{-is} \right)$$

we can evaluate the integral. If we then use (7.6) again and and apply the simple estimate

$$\Big| \pm 1 - \frac{1}{\varepsilon} \Big| \ge 1 \quad \text{for} \quad 0 < \varepsilon \le \frac{1}{2}$$

we find that

$$\Big| \int_0^t e^{-is/\varepsilon} \sin s \, ds \Big| \le 2 \ .$$

This yields the bound $|\delta(t)| \le 4\varepsilon$ for all t. ◇

Extension: Using the notations of the above proof, we have $u(x,t) = \sin t + e^{it/\varepsilon} + \delta(t,\varepsilon)$ where $\delta' = \frac{i}{\varepsilon} \delta - \cos t, \delta(0,\varepsilon) = 0$. One can now modify the two steps that led to the $\mathcal{O}(\varepsilon)$–approximation of $u(x,t)$ and approximate δ. In this way one obtains an $\mathcal{O}(\varepsilon^2)$ approximation for $u(t,\varepsilon)$. This process can be continued to any order in ε.

For many extensions of the simple example discussed here we refer to [10].

3. A Difficult Example

Suppose we consider two planets moving around the sun. For definiteness, let's consider Earth and Jupiter and disregard all other planets. Suppose, then, we first ignore the gravitational attraction between Earth and Jupiter and only take the interaction between the planets and the sun into account. In this simplified model, both planets move around the sun in elliptical orbits, as we have studied in Chapter 4. Each planet moves periodically.

Think of this periodic motion as the *fast* motion. If we now refine our model and take the interaction between Earth and Jupiter into account, then the orbits of both planets will *slowly* change. We have evolution on two different time scales: The *fast* motion of the planets in their orbits and the much *slower* change of the orbits themselves.

As a small parameter in the refined model one can identify the quotient

$$\varepsilon = \frac{mass_{Jupiter}}{mass_{sun}} \sim 0.001 \; .$$

What is the separation of time scales between the orbital motion and the motion which changes the orbits? It appears that the Earth's orbit did not change much during the past 10^3 or even 10^6 years. This indicates that the separation of time scales is neither $1/\varepsilon$ nor $1/\varepsilon^2$. The analysis of this example is very difficult.

Chapter 8

Stability and Bifurcations

Summary: In this chapter we introduce the important concept of *asymptotic stability* of fixed points and then consider two simple growth models: exponential and logistic growth. For both models, the trivial fixed point $u^* = 0$ is unstable, but logistic growth also has an asymptotically stable fixed point.

Discretizing time in the logistic model, we will derive the so–called delayed logistic map. As a parameter λ increases, the fixed point P_λ loses its stability in a Neimark–Sacker (or second Hopf) bifurcation : an attracting invariant circle is born and bifurcates from the branch of fixed point. The analysis of this bifurcation is quite difficult and will not be discussed here. [1] Instead, we demonstrate it numerically. As the parameter λ increases further, the invariant circle disappears in a complicated fashion.

Turning back to the ODE case, we will use the equation

$$u' = u(1+u)(1-u) + \lambda \tag{8.1}$$

and the stability of its fixed points to illustrate the phenomenon of *hysteresis*.

1. Fixed Points

Consider the initial value problem

$$u' = f(u), \quad u(0) = u_0 \ , \tag{8.2}$$

where $f : \mathbb{R}^N \to \mathbb{R}^N$ is a C^1–function. Denote the solution by $u(t; u_0)$. A point $u^* \in \mathbb{R}^N$ is called a fixed point of the evolution if $f(u^*) = 0$ because $u(t; u^*) \equiv u^*$ if and only if $f(u^*) = 0$. Roughly speaking, a fixed point u^* is called asymptotically stable if all initial states u_0 close to u^* evolve towards u^*.

The technical definition is more complicated:

[1] The book of Kuznetsov [12] gives a careful analysis of this bifurcation.

Definition 8.1 *A fixed point u^* is called asymptotically stable if the following two conditions hold:*

(a) For all $\varepsilon > 0$ there is $\delta > 0$ so that $|u^* - u_0| < \delta$ implies $|u^* - u(t; u_0)| < \varepsilon$ for all $t \geq 0$.

(b) There is $\delta > 0$ so that $|u^* - u_0| < \delta$ implies $u(t; u_0) \to u^*$ as $t \to \infty$.

If (a) holds then u^* is called stable . If (a) does not hold, then u^* is called unstable.

We will use the following result:

Theorem 8.1. *Denote by $A = f'(u^*)$ the Jacobian of f at the fixed point u^*. If all eigenvalues λ_j of A satisfy $\text{Re}\,\lambda_j < 0$ then u^* is asymptotically stable. If A has an eigenvalue λ_j with $\text{Re}\,\lambda_j > 0$ then u^* is unstable.*

For a proof, see Chapter 13 of [3], for example.[2]

If the dimension of the state space $\mathbb{R}^N$ is $N = 1$ then the Jacobian A is just a number, the derivative of the function f at the fixed point u^*. As an example, consider the scalar equation

$$u' = u(1+u)(1-u) =: f(u) \ . \tag{8.3}$$

The three zeros of f are the fixed points. Since

$$f'(\pm 1) < 0 < f'(0)$$

the fixed points $u^*_{1,2} = \pm 1$ are asymptotically stable and $u^*_3 = 0$ is unstable. This is also obvious if one draws the phase line for $u(t)$.

In Section 5., we will add a parameter λ to the cubic $f(u)$ in (8.3) and consider equation (8.1). As λ changes, the fixed points move and may disappear. We will use equation (8.1) to show the phenomenon of hysteresis.

Fixed points and their stability are also important for discrete–time evolutions $u_{n+1} = \Phi(u_n)$ determined by a map $\Phi : \mathbb{R}^N \to \mathbb{R}^N$. A fixed point $u^* \in \mathbb{R}^N$ is a point with $\Phi(u^*) = u^*$. Again, roughly speaking, u^* is called asymptotically stable if all initial states u_0 close to u^* evolve towards u^*. We must modify Definition 8.1 only slightly:

Definition 8.2 *A fixed point u^* of Φ is called asymptotically stable if the following two conditions hold:*

(a) For all $\varepsilon > 0$ there is $\delta > 0$ so that $|u^* - u_0| < \delta$ implies $|u^* - \Phi^n(u_0)| < \varepsilon$ for all $n \geq 0$.

(b) There is $\delta > 0$ so that $|u^* - u_0| < \delta$ implies $\Phi^n(u_0) \to u^*$ as $n \to \infty$.

If (a) holds then u^* is called stable. If (a) does not hold, then u^* is called unstable.

[2]It is an interesting task to *quantify* this stability result. For example, how does one determine a *realistic* value for $\delta > 0$ so that $|u^* - u_0| < \delta$ implies $u(t, u_0) \to u^*$ as $t \to \infty$? For some results about this question, see [11].

As is the case of a differential equation, one can often establishes stability or instability of a fixed point, by looking at the eigenvalues of the Jacobian $A = \Phi'(u^*)$. Instead of Theorem 8.1 the following holds in the map–case. It is assumed that $\Phi : \mathbb{R}^N \to \mathbb{R}^N$ is C^1.

Theorem 8.2. *Denote by $A = \Phi'(u^*)$ the Jacobian of Φ at the fixed point u^*. If all eigenvalues λ_j of A satisfy $|\lambda_j| < 1$ then u^* is asymptotically stable. If A has an eigenvalue λ_j with $|\lambda_j| > 1$ then u^* is unstable.*

In Section (4.) we will apply this theorem to the delayed logistic map.

2. Exponential Growth

The simplest growth model is given by the ODE initial–value problem

$$u'(t) = bu(t), \quad u(0) = u_0 \ .$$

Here b is the birth rate per individual, which we assume to be positive, and u_0 is the population size at time $t = 0$. The solution is $u(t) = u_0 e^{bt}$. A standard derivation of the equation $u' = bu$ proceeds as follows: If $u(t)$ is the population size at time t and the birth rate per individual in the time interval $[t, t + \Delta t]$ is, approximately,

$$b\Delta t$$

then

$$u(t + \Delta t) \sim u(t) + bu(t)\Delta t \ .$$

Subtract $u(t)$ from both sides, divide by Δt, and let $\Delta t \to 0$ to obtain $u' = bu$.

As a simple practice, let us apply Theorem 8.1 to the equation $u' = bu$ where $b > 0$. Here $f(u) = bu$ has the unique zero $u^* = 0$. Since $f'(0) = b > 0$, the fixed point $u^* = 0$ is unstable.

If $b_1 > 0$ is the birth rate and $d_1 > 0$ the death rate, then the above arguments lead to the equation

$$u' = bu \quad \text{with} \quad b = b_1 - d_1 \ .$$

If $d_1 > b_1$ then $b < 0$ and the fixed point $u^* = 0$ becomes asymptotically stable.

3. Logistic Growth

Exponential growth cannot continue for long.[3] A somewhat more realistic model is so–called logistic growth given by

[3] An investment of one cent in the Middle Kingdom of Egypt, 4,000 years ago, with an interest rate of 2%, continuously compounded, would have grown to $(e^{0.02})^{4,000}$ cents. That is about 100 trillion dollars on

$$u'(t) = r\Big(1 - \frac{u(t)}{K}\Big)u(t) \tag{8.4}$$

where r and K are positive constants. Here the birth rate b is replaced by

$$b \sim r\Big(1 - \frac{u(t)}{K}\Big) \; .$$

The constant K is the carrying capacity of the system. If $u(t) << K$ then $u' \sim ru$, and we have approximately exponential growth. As $u(t)$ approaches K, the growth rate decays to zero.

To apply Theorem 8.1, we note that the function

$$f(u) = r(1 - \frac{u}{K})u$$

has the zeros $u_1^* = 0$ and $u_2^* = K$. The fixed point $u_1^* = 0$ is unstable, whereas $u_2^* = K$, the carrying capacity, is asymptotically stable.

One can confirm this by solving the logistic equation explicitly. Applying separation of variables, one obtains the general solution

$$u(t) = \frac{K}{1 + \gamma e^{-rt}} \; ,$$

depending on the free parameter γ. The initial condition $u(0) = u_0$ yields

$$\gamma = \frac{K - u_0}{u_0} \; ,$$

thus

$$u(t) = \frac{K u_0}{u_0 + (K - u_0)e^{-rt}} \; . \tag{8.5}$$

The following is left as an exercise: In the above formula for $u(t)$ assume that $r > 0$ and $K > 0$. Show:

(a) If $u_0 > 0$ then $u(t) \to K$ as $t \to \infty$.

(b) If $u_0 = 0$ then $u(t) \equiv 0$.

(c) If $u_0 < 0$ then $u(t) \to -\infty$ as $t \to t^*-$ where $t^* > 0$ is the time at which the denominator in (8.5) becomes zero.

ODEs that are somewhat more complicated than (8.4) are often difficult or impossible to solve analytically. One then relies on numerical methods. Matlab provides well–tested tools, the most basic one is *ode45*. In the scripts listed in Section 1. we use *ode45* to solve the logistic equation $u' = u(1 - u)$ with ten different initial values. The solutions are plotted in Figure 8.1.

---

every square centimeter of the surface of the earth. We wish the economy would always grow at least by a few percent every years, but that's impossible in the long run.

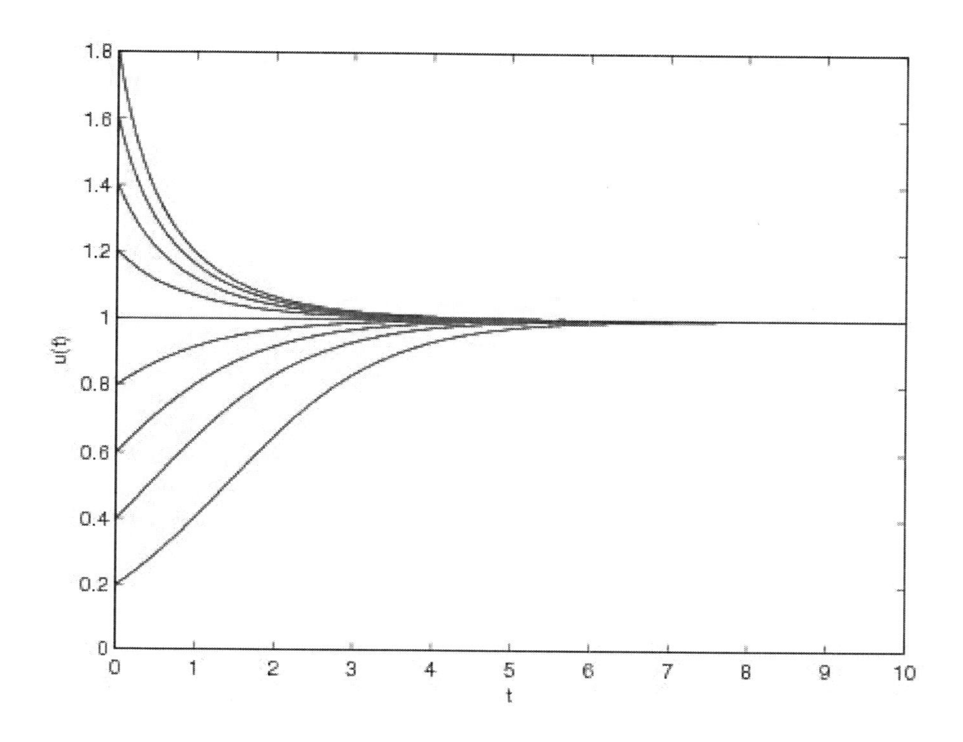

Figure 8.1. Solutions of $u' = u(1-u)$ for initial conditions $u(0) = 0, 0.2, 0.4, ..., 1.8$

4. The Delayed Logistic Map

If one replaces the logistic differential equation $u' = r(1 - u/K)u$ by the discrete–time model

$$\frac{1}{\Delta t}(u_{n+1} - u_n) = r(1 - u_{n-1}/K)u_n, \quad n = 1, 2, \ldots \tag{8.6}$$

then one obtains evolution by the so–called delayed logistic map. The term *delayed* is used since the growth coefficient $r(1 - u/K)$ is evaluated at u_{n-1} instead of u_n. The dynamics (8.6) can be visualized by a sequence of points (u_{n-1}, u_n) in the plane. In this section, we will show that the dynamics (8.6) is qualitatively more complicated, in general, than the dynamics of the logistic differential equation. The latter is qualitatively determined by its two fixed points, $u_1^* = 0$ (unstable) and $u_2^* = K$ (stable).

Reduction to Dependency on One Parameter: Equation (8.6) contains three positive parameters, $\Delta t, r$, and K. Let us show first that, by proper scaling of u_n, the dynamics depends essentially on one parameter only. We write

$$\begin{aligned} u_{n+1} &= u_n + r\Delta t(1 - u_{n-1}/K)u_n \\ &= u_n\Big(1 + r\Delta t - r\Delta t u_{n-1}/K\Big) \\ &= (1 + r\Delta t)u_n(1 - \beta u_{n-1}) \end{aligned}$$

with

$$\beta = \frac{r\Delta t}{(1 + r\Delta t)K} \ .$$

Set

$$\lambda = 1 + r\Delta t \ . \qquad (8.7)$$

Then the discrete–time evolution (8.6) is equivalent to

$$u_{n+1} = \lambda u_n(1 - \beta u_{n-1}), \quad n = 1, 2, \ldots$$

With a constant q (independent of n), to be determined below, set

$$u_n = qv_n$$

for all n. Obtain

$$qv_{n+1} = \lambda qv_n(1 - \beta qv_{n-1}) \ .$$

Choosing $q = 1/\beta$ we arrive at

$$v_{n+1} = \lambda v_n(1 - v_{n-1}), \quad n = 1, 2, \ldots \qquad (8.8)$$

Equation (8.8) is called evolution by the delayed logistic map with parameter λ. If we set

$$F_\lambda(x, y) = \Big(y, \lambda y(1 - x)\Big) \quad \text{for} \quad (x, y) \in \mathbb{R}^2 \qquad (8.9)$$

then the evolution (8.8) becomes

$$(v_{n-1}, v_n) \to F_\lambda(v_{n-1}, v_n) = (v_n, v_{n+1}) \ .$$

In other words, we must iterate the map F_λ.

Fixed Points and Their Stability: Since $\lambda = 1 + r\Delta t$ (see (8.7)) we will assume $\lambda > 1$ in the following. Let us determine the fixed points of F_λ: The equation $F_\lambda(x, y) = (x, y)$ is equivalent to

$$x = y \quad \text{and} \quad y = \lambda y(1 - x) \ .$$

There is the trivial fixed point $Q = (0,0)$ and the nontrivial fixed point

$$P_\lambda = \Big(1 - \frac{1}{\lambda}, 1 - \frac{1}{\lambda}\Big) .$$

We will apply Theorem 8.2 to discuss their stability. The Jacobian of F_λ is

$$F'_\lambda(x,y) = \begin{pmatrix} 0 & 1 \\ -\lambda y & \lambda(1-x) \end{pmatrix}$$

and, therefore,

$$F'_\lambda(0,0) = \begin{pmatrix} 0 & 1 \\ 0 & \lambda \end{pmatrix} .$$

Since $\lambda > 1$ the trivial fixed point $Q = (0,0)$ is unstable. (Recall that $u_1^* = 0$ is an unstable fixed point of the logistic equation. Thus, regarding the unstable fixed point, the continuous–time and the discrete–time models are in agreement.)

The Jacobian of F_λ at P_λ is

$$F'_\lambda(P_\lambda) = \begin{pmatrix} 0 & 1 \\ 1-\lambda & 1 \end{pmatrix} .$$

The eigenvalues $\mu_{1,2}$ are:

$$\mu_{1,2} = \begin{cases} \frac{1}{2} \pm \sqrt{\frac{5}{4} - \lambda} & \text{for } 1 < \lambda \le \frac{5}{4} , \\ \frac{1}{2} \pm i\sqrt{\lambda - \frac{5}{4}} & \text{for } \lambda > \frac{5}{4} . \end{cases}$$

It is elementary to show that $|\mu_{1,2}| < 1$ for $1 < \lambda \le \frac{5}{4}$ and

$$|\mu_{1,2}|^2 = \lambda - 1 \quad \text{for} \quad \lambda > \frac{5}{4} .$$

Thus, as long as $1 < \lambda < 2$, the eigenvalues $\mu_{1,2}$ of the Jacobian $F'_\lambda(P_\lambda)$ lie strictly inside the unit disk. By Theorem 8.2, the fixed point P_λ is asymptotically stable for $1 < \lambda < 2$.

If $\lambda = 2$ then the eigenvalues

$$\mu_{1,2} = \frac{1}{2} \pm i\frac{\sqrt{3}}{2} \tag{8.10}$$

lie on the unit circle and if $\lambda > 2$ they lie strictly outside the unit disk. Therefore, by Theorem 8.2, the fixed point P_λ is unstable for $\lambda > 2$. (If $\lambda = 2$ then Theorem 8.2 gives no information about stability or instability of the fixed point P_2.)

Let us summarize: For $\frac{5}{4} < \lambda < \infty$ the eigenvalues $\mu_{1,2} = \mu_{1,2}(\lambda)$ of the Jacobian $F'_\lambda(P_\lambda)$ form a complex conjugate pair. It lies strictly inside the unit disk for $\frac{5}{4} < \lambda < 2$ and strictly outside for $\lambda > 2$. As λ crosses from $\lambda < 2$ to $\lambda > 2$ the pair $\mu_{1,2}(\lambda)$ leaves the unit disk and the fixed point F_λ loses its stability.

Neimark–Sacker Bifurcation:

A general question of bifurcation theory can be vaguely formulated as follows: Consider an evolution equation $u_{n+1} = \Phi(u_n, \lambda)$ or $u' = f(u, \lambda)$, depending on a real parameter λ. Assume a *phenomenon* (e.g. a fixed point) is stable for $\lambda < \lambda^*$, but unstable for $\lambda > \lambda^*$. Then, as the parameter crosses from $\lambda < \lambda^*$ to $\lambda > \lambda^*$, which *new phenomenon* (if any) will take over stability?

For the maps F_λ (see (8.9)) the fixed point P_λ loses stability as λ crosses from $\lambda < 2$ to $\lambda > 2$. This happens because a pair of complex conjugate eigenvalues of $F'_\lambda(P_\lambda)$ leaves the unit circle. A general result of bifurcation theory, the Neimark–Sacker Theorem , deals with this kind of loss of stability and, under additional assumptions, shows that an attracting invariant curve replaces the unstable fixed point. The theorem is difficult to prove (see, for example, Chapter 4.7 of [12]), in general, but we try to make it plausible for the family of maps (8.9).

As often, *linearization* gives a first hint at what happens. Let us consider the map F_λ for $\lambda = 2$ and linearize the dynamics about the fixed point P_2. We have

$$F_2(P_2) = P_2, \quad F_2'(P_2) = \begin{pmatrix} 0 & 1 \\ -1 & 1 \end{pmatrix} =: A\ .$$

Therefore,

$$\begin{aligned} F_2(P_2 + \varepsilon(x, y)) &= F_2(P_2) + \varepsilon A \begin{pmatrix} x \\ y \end{pmatrix} + \mathcal{O}(\varepsilon^2) \\ &= P_2 + \varepsilon A \begin{pmatrix} x \\ y \end{pmatrix} + \mathcal{O}(\varepsilon^2) \end{aligned}$$

Neglecting the $\mathcal{O}(\varepsilon^2)$–term, we see that the evolution of the deviations from P_2 is governed by the linear map

$$\begin{pmatrix} x \\ y \end{pmatrix} \to A \begin{pmatrix} x \\ y \end{pmatrix} . \tag{8.11}$$

Here the matrix A has its eigenvalues $\mu_{1,2}$ on the unit circle. (See (8.10).) Since

$$S^{-1}AS = \begin{pmatrix} \mu_1 & 0 \\ 0 & \mu_2 \end{pmatrix} =: D \quad \text{with} \quad S = \begin{pmatrix} 1 & 1 \\ \mu_1 & \mu_2 \end{pmatrix} \quad \text{and} \quad |\mu_j| = 1$$

and [4]

$$(S^{-1})^* S^{-1} = \frac{1}{|detS|^2} \begin{pmatrix} 2 & -1 \\ -1 & 2 \end{pmatrix}$$

[4] If $B = (b_{jk}) \in \mathbb{C}^{N \times N}$ is a matrix, we denote its adjoint by $B^* = (\bar{b}_{kj})$. An important rule is $\langle Bu, v\rangle = \langle u, B^*v\rangle$ for all vectors $u, v \in \mathbb{C}^N$.

it is not difficult to show that the linear map (8.11) leaves every ellipse of the family

$$(x, y)\begin{pmatrix} 2 & -1 \\ -1 & 2 \end{pmatrix}\begin{pmatrix} x \\ y \end{pmatrix} = const = \alpha \tag{8.12}$$

invariant. Precisely: For every $\alpha > 0$ the points $(x, y) \in \mathbb{R}^2$ satisfying (8.12) form an ellipse, E_α. Since

$$\left| S^{-1} A \begin{pmatrix} x \\ y \end{pmatrix} \right| = \left| S^{-1} \begin{pmatrix} x \\ y \end{pmatrix} \right|$$

the map (8.11) maps the ellipse E_α onto itself. Setting

$$u = \begin{pmatrix} x \\ y \end{pmatrix}$$

the above equation follows from

$$\begin{aligned} \langle S^{-1}Au, S^{-1}Au \rangle &= \langle S^{-1}ASS^{-1}u, S^{-1}ASS^{-1}u \rangle \\ &= \langle DS^{-1}u, DS^{-1}u \rangle \\ &= \langle S^{-1}u, S^{-1}u \rangle \end{aligned}$$

The last equation holds since the diagonal entries μ_j of D have absolute value 1.

This analysis of the linearized problem at $\lambda = 2$ gives a hint at what happens for the nonlinear problem for $2 < \lambda < 2 + \varepsilon$: One can show that the map $(x, y) \to F_\lambda(x, y)$ has an *invariant curve* Γ_λ which, approximately, is an ellipse surrounding P_λ. This follows from the Neimark–Sacker bifurcation theorem.

To illustrate the dynamics of the delayed logistic map, consider first the value $\lambda = 1.9$. For this value of λ, the fixed point

$$P_{1.9} = \left(1 - \frac{1}{1.9}, 1 - \frac{1}{1.9}\right) = (0.4737, 0.4737)$$

is asymptotically stable. If we start the evolution at the point $(0.5, 0.6)$ we get a sequence of points converging to $P_{1.9}$. This is shown in Figures 8.2. The scripts progtral.m and progtra2.m, listed in Section 2., have been used. If one changes the value of λ to $\lambda = 2.1$ then the fixed point P_λ is unstable. The output of the computation is shown in Figures 8.3.

To show the invariant curve for $\lambda = 2.1$ more clearly, run the code of script progtra5.m listed in Chapter 2.. The resulting curve is shown in Figure 8.4(a).

If one increases λ further, the invariant curve eventually breaks and disappears in a complicated fashion, which is not well understood. For example, using the code progtra5.m with $\lambda = 2.27$ produces Figure 8.4(b). The invariant curve is broken.

5. Parameter Dependent Evolution and Hysteresis

In this section we will illustrate the important phenomenon of hysteresis for a parameter dependent evolution. As a simple example, consider the ODE

$$u' = u(1+u)(1-u) + \lambda =: f(u,\lambda) \tag{8.13}$$

where λ is a real parameter. In Figure 8.5 we graph the polynomial $u \to f(u,0) = u(1+u)(1-u)$ with its three zeros, $u^*_{1,3} = \pm 1$ and $u^*_2 = 0$. These are the fixed points of (8.13) for $\lambda = 0$. We know from Theorem 8.1 that the fixed points $u^*_{1,3}$ are asymptotically stable, whereas $u^*_2 = 0$ is unstable.

As λ changes, the fixed points become functions of λ and we write $u^*_j(\lambda)$ for the zeros of $f(\cdot,\lambda)$. It is easy to check that the local maximum in Figure 8.5 is $M = 2/3\sqrt{3}$ and the local minimum is $-M$. If we start at $\lambda = 0$ and then increase λ in the interval $0 \le \lambda \le M$, the fixed points $u^*_1(\lambda)$ and $u^*_2(\lambda)$ approach each other and collide for $\lambda = M$. In the same way, if we start at $\lambda = 0$ and then decrease λ in the interval $-M \le \lambda \le 0$, the fixed points $u^*_2(\lambda)$ and $u^*_3(\lambda)$ approach each other and collide for $\lambda = -M$.

In Figure 8.6 we graph all the fixed points $u^*_{1,2,3}(\lambda)$. The three branches form a single curve. Note that the branches $u^*_1(\lambda)$ and $u^*_3(\lambda)$ consist of asymptotically stable fixed points whereas the middle branch $u^*_2(\lambda)$ consists of unstable fixed points.

Let us now discuss what all this implies for the time dependent ODE (8.13). Suppose we make the parameter λ in ODE (8.13) a slowly varying function of time. To be specific, let us take

$$\lambda(t) = \sin(2\pi t/100) \quad \text{for} \quad 0 \le t \le 200$$

and let us use the initial condition $u(0) = 1$. Figure 8.7 shows the function $\lambda(t)$, going through two periods of the sine–function, and also shows the (numerically computed) solution $u(t)$.

The point is this: Since $\lambda(t)$ changes rather slowly, the solution $u(t)$ is always close to a stable fixed point of the equation (8.13), where the fixed point is determined *as if λ is frozen at time t*. At time $t = 0$ we have $\lambda(0) = 0$ and the solution starts at the stable fixed point $u^*_3(0) = 1 = u(0)$. Now consider the time interval $0 \le t \le 50$. The function $\lambda(t)$ increases from zero to one, and then decreases from one to zero. The solution $u(t)$ follows the corresponding fixed point $u^*_3(\lambda)$ and we have, approximately, $u(t) \sim u^*_3(\lambda(t))$ for $0 \le t \le 50$.

But now let's go through the next half–period: As time changes in the interval $50 \leq t \leq 100$, the parameter $\lambda(t)$ goes from zero to minus one and then back to zero. When $\lambda(t)$ decreases below $-M$, the fixed point $u_3^*(\lambda)$ no longer exists! The solution $u(t)$ then moves, rather rapidly, to the other stable fixed point, $u_1^*(\lambda(t))$.

Look at Figure 8.7 at time $t = 100$. The parameter $\lambda(t)$ has returned to its initial value, $\lambda(100) = \lambda(0) = 0$. However, for the solution $u(t)$, always following a stable fixed point, we have $u(100) \sim -1$, but started at $u(0) = 1$. In other words, though the parameter $\lambda(t)$ has returned to its starting value, the solution $u(t)$ has moved from $u(0) = u_3^*(0) = 1$ to $u(100) \sim u_1^*(0) = -1$. It has moved from one stable fixed point to the other.

Using Figure 8.7, the reader may check what happens in the next period, $100 \leq t \leq 200$.

In Figure 8.8) we graph all points $(\lambda(t), u(t))$ which occur during the evolution sketched in Figure 8.7. We see a so–called hysteresis loop. The word hysteresis is derived from an ancient Greek word and means something like *lagging behind.* The state of a system, described by $u(t)$, not only depend on the current parameter value $\lambda(t)$ (the current environment), but also depends on the past. If $\lambda(t)$ changes periodically, then one can expect that $u(t)$ will approach periodic behavior, but there is a time lag. More importantly, if $\lambda(t)$ lies in the interval $-M < \lambda < M$, then $u(t)$ is either close to $u_1^*(\lambda(t))$ or to $u_3^*(\lambda(t))$. Which of these two values is approximated depends on the past.

Discussion: In applications, evolution equations always depend on parameters. To mention a difficult, but interesting, example, think of equations for weather/climate and the CO_2–concentration in the atmosphere as a parameter. As a first approximation, one may fix parameter values and obtain a solution in a certain regime. If the parameters change slowly, the solution will typically adjust slowly, and small changes in the parameters will typically lead to small changes in the solution. However, bifurcations may occur. A solution may lose stability as a parameter changes. (In equation (8.13), bifurcations occur if λ changes through $\pm M = \pm 2/3\sqrt{3}$.)

If one does not like a new solution and forces the parameters back to their old values, there is no guarantee that the solution will return to its old regime. Instead, it may follow a new branch. To come back to the (speculative) CO_2–example: We are increasing the CO_2–concentration in the atmosphere. Suppose we do not like the new resulting climate and force the concentration back to its pre–industrial level. Will we get our old climate back? The climate may remain in a new regime.

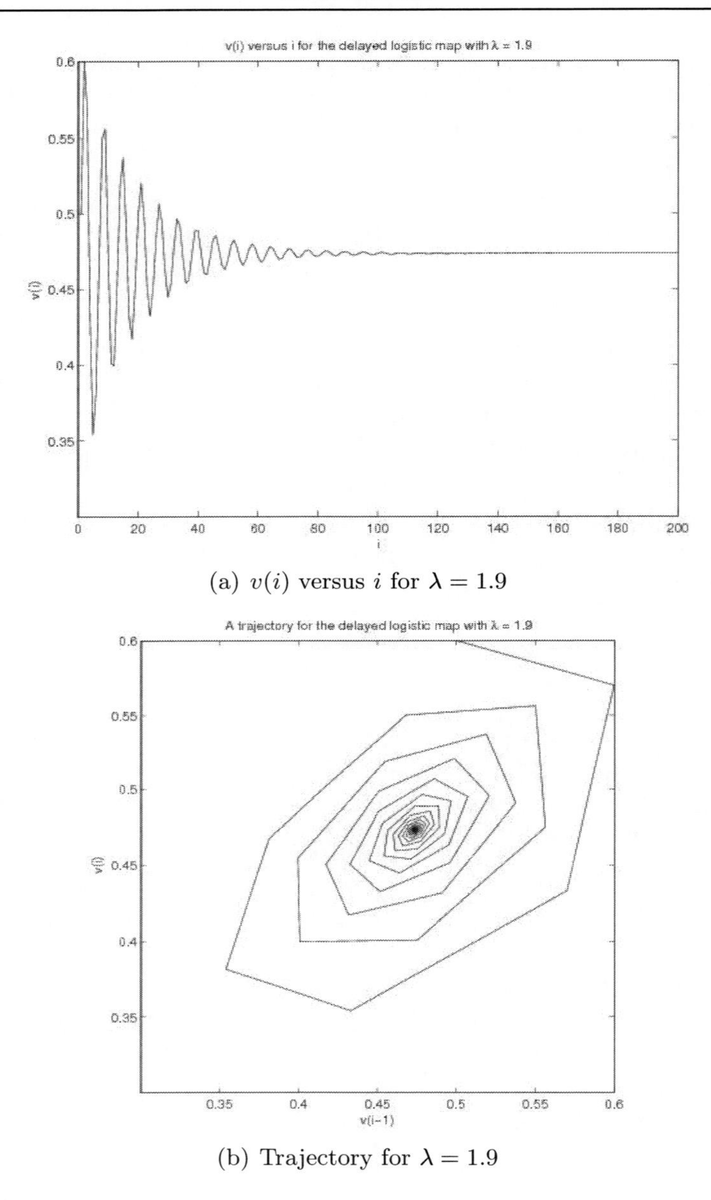

(a) $v(i)$ versus i for $\lambda = 1.9$

(b) Trajectory for $\lambda = 1.9$

Figure 8.2. Evolution of the Delayed Logistic Map for $\lambda = 1.9$

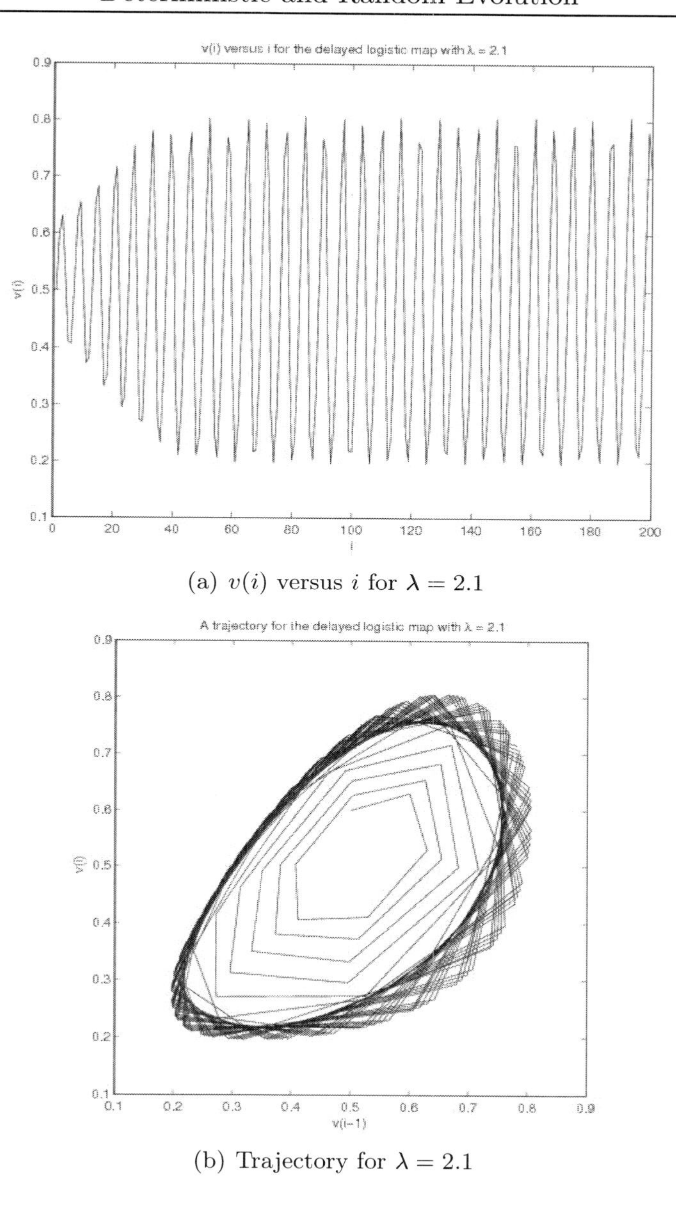

(a) $v(i)$ versus i for $\lambda = 2.1$

(b) Trajectory for $\lambda = 2.1$

Figure 8.3. Evolution of the Delayed Logistic Map for $\lambda = 2.1$

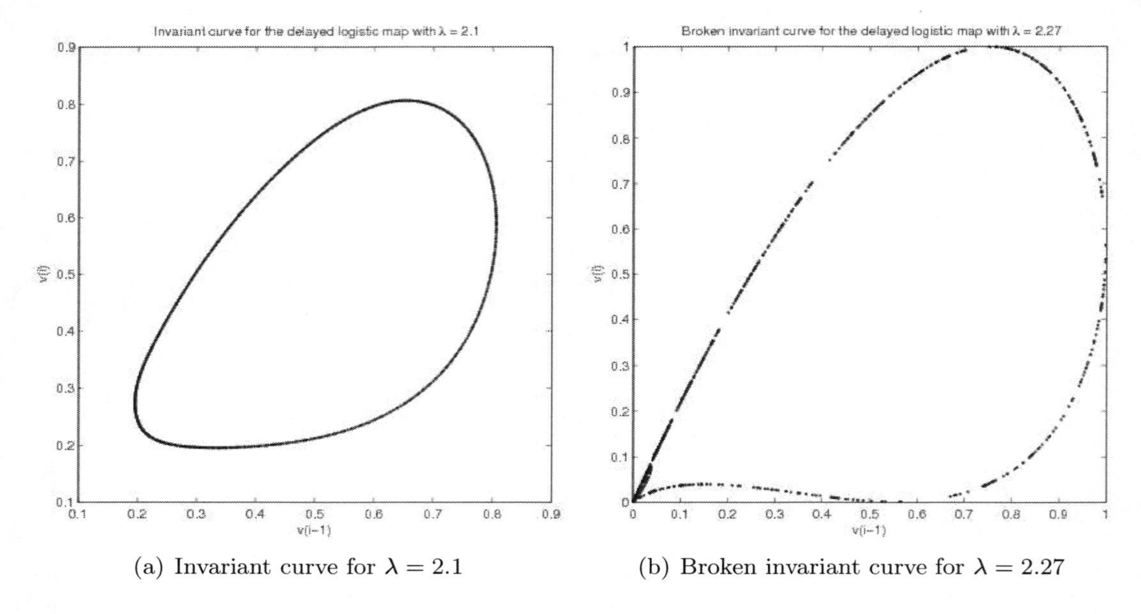

(a) Invariant curve for $\lambda = 2.1$ (b) Broken invariant curve for $\lambda = 2.27$

Figure 8.4. Breakdown of Invariant Curve for the Delayed Logistic Map

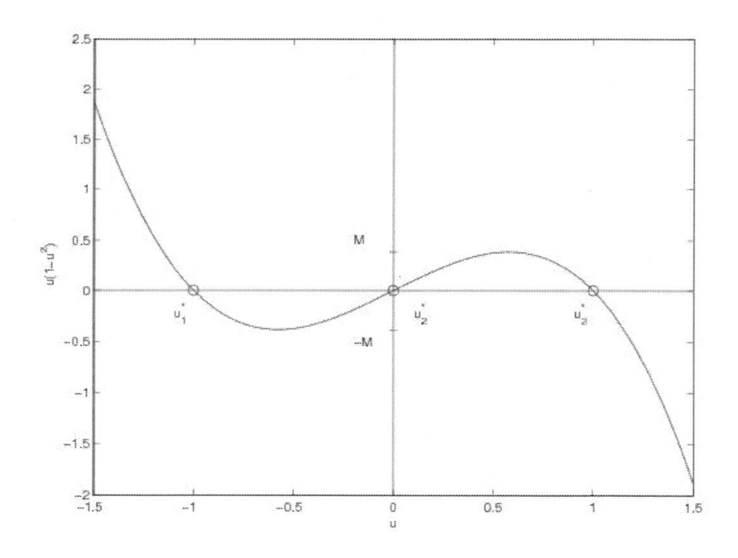

Figure 8.5. Polynomial with zeros u_j^* and local extrema $\pm M$

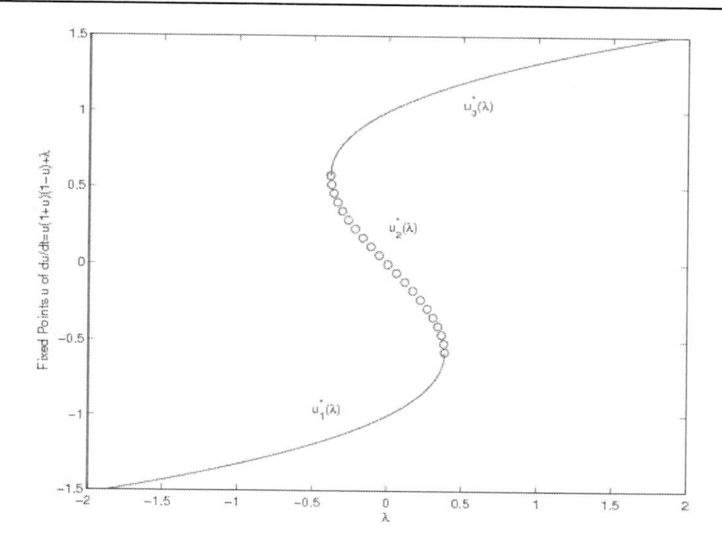

Figure 8.6. Hysteresis: Branch of fixed points

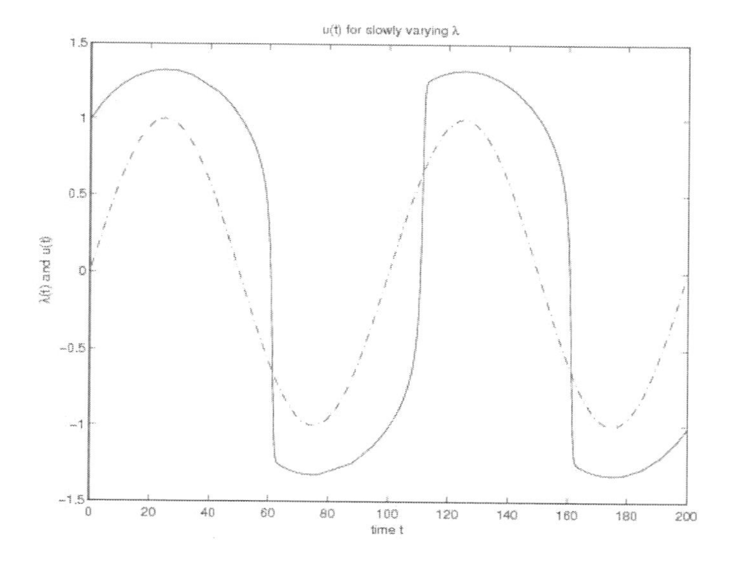

Figure 8.7. $\sin(2\pi t/100)$ and $u(t)$

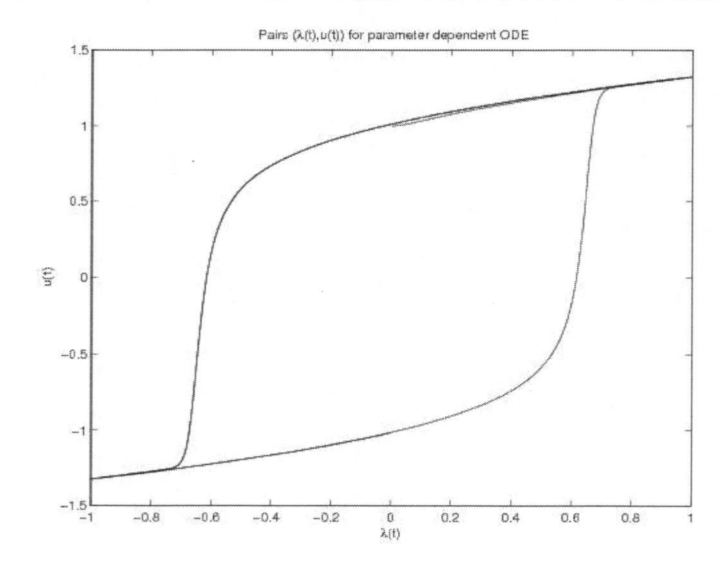

Figure 8.8. Hysteresis loop for ODE (8.13)

Chapter 9

Scripts

Summary: This chapter contains the Matlab scripts that were used to generate the figures of the previous chapter. The codes are neither elegant nor efficient, but hopefully easy to read.

For readers who are completely unfamiliar with Matlab: A line starting with % is just a comment.

The Matlab subroutine ode45 is a basic tool for solving an initial value problem for an ODE numerically. In the script of Section 1. we show how to solve the logistic equation $u' = u(1-u)$ with ode45. The scripts *p3.m* and *p4.m* in Section 3. use the Matlab subroutine *ode15s* instead of *ode45*, because the differential equation is stiff.

1. Script for Logistic Growth

```
%%%%%%%%%%%%%%%%%%%%%%%%%%%%%%%%%%%%%%%%%%%%%%%%%%%%%
%
% This script solves the ODE u'=u(1-u).
%
%%%%%%%%%%%%%%%%%%%%%%%%%%%%%%%%%%%%%%%%%%%%%%%%%%%%%

% definition of the RHS of the ODE y'=f(t,y)
% change f here if needed
f=@(t,y) y*(1-y);

% loop for different initial values
for i=1:10
    % solving ODE numerically with ODE45
    % here f is the RHS of y'=f, it has to be function header
    % [0,10] is the interval where we compute the solution
```

```
    % [(i-1)*.2] is the initial value, i.e., y(0)
    [T,Y] = ode45(f,[0 10],[(i-1)*.2]);
    % plotting all the curves in one plot
    % 'hold on' keeps' the same plot for all initial values
    plot(T,Y,'-b'); hold on;
end;
```

2. Scripts for the Delayed Logistic Map

```
%progtra1.m
% iteration of the map
% F_\lambda(x,y)=(y,\lambda y (1-x))
% starting at (x,y)=(0.5,0.6).
%%%%%%%%%%%%%%%%%%%%%%%%%%%%%%%%%%%%%%%%
% plot of v(i) versus i
%%%%%%%%%%%%%%%%%%%%%%%%%%%%%%%%%%%%%%%%
% convergence to a fixed point
%%%%%%%%%%%%%%%%%%%%%%%%%%%%%%%%%%%%%%%%
lam=1.9;
clear v;
v(1)=0.5;
v(2)=0.6;
for i=3:200
v(i)=lam*v(i-1)*(1-v(i-2));
end;
plot(v)
xlabel('i')
ylabel('v(i)')
title('v(i) versus i for the delayed logistic map with \lambda = 1.9')

%progtra2.m
% iteration of the map
% F_\lambda(x,y)=(y,\lambda y (1-x))
% starting at (x,y)=(0.5,0.6).
%%%%%%%%%%%%%%%%%%%%%%%%%%%%%%%%%%%%%%%%%
% plot of the points (v(i),v(i+1))
% in the phase plane
%%%%%%%%%%%%%%%%%%%%%%%%%%%%%%%%%%%%%%%%%
% convergence to a fixed point
%%%%%%%%%%%%%%%%%%%%%%%%%%%%%%%%%%%%%%%%%
lam=1.9;
```

```
clear v;
v(1)=0.5;
v(2)=0.6;
for i=3:200
v(i)=lam*v(i-1)*(1-v(i-2));
end;
plot(v(1:199),v(2:200))
axis square
xlabel('v(i-1)')
ylabel('v(i)')
title('A trajectory for the delayed logistic map with \lambda = 1.9')

%progtra5.m
% iteration of the map
% F_\lambda(x,y)=(y,\lambda y (1-x))
% starting at (x,y)=(0.5,0.6).
%%%%%%%%%%%%%%%%%%%%%%%%%%%%%%%%%%%%%%%%%%
% plot of the points (v(i),v(i+1))
% in the phase plane
%%%%%%%%%%%%%%%%%%%%%%%%%%%%%%%%%%%%%%%%%%
% convergence to an invariant curve
%%%%%%%%%%%%%%%%%%%%%%%%%%%%%%%%%%%%%%%%%%
lam=2.1;
clear v;
v(1)=0.5;
v(2)=0.6;
for i=3:1000
v(i)=lam*v(i-1)*(1-v(i-2));
end;
plot(v(200:999),v(201:1000),'.')
axis square
xlabel('v(i-1)')
ylabel('v(i)')
title('Invariant curve for the delayed logistic map with \lambda = 2.1')
```

3. Scripts for Parameter Dependent Evolution and Hysteresis

The following Matlab script generates Figure 8.5.

```
%%%%%%%%%%%%%%%%%%%%%%%%%%%%%%%%%%%%%%%%%%
```

```
% p2.m: polynomial u(1+u)(1-u)
%%%%%%%%%%%%%%%%%%%%%%%%%%%%%%%%%%%%%%%%%%%%
clear u*
N=101;
h=0.2;
u=linspace(-1.5,1.5,N);
v=linspace(-2,2.5,N);
for j=1:N
p(j)=u(j)*(1+u(j))*(1-u(j));
zero(j)=0;
end
plot(u,p,u,zero)
%axis equal
hold on
plot(zero,v)
hold on
text(-1.1, -0.3, 'u_1^*')
text(0.1, -0.3, 'u_2^*')
text(0.9, -0.3, 'u_3^*')
text(-0.2, 0.5, 'M')
text(-0.2, -0.5, '-M')
xlabel('u')
ylabel('u(1-u^2)')
%%%%%%%%%%%%%%%%%%%%%%%%%%%%%%%%%%%%%%%%%%%%%%%%%%
% circles around the zeros
%%%%%%%%%%%%%%%%%%%%%%%%%%%%%%%%%%%%%%%%%%%%%%%%%%
rs=0.05;
rc=0.025;
t=linspace(0,2*pi,N);
for j=1:N
s(j)=rs*sin(t(j));
c(j)=rc*cos(t(j));
cp(j)=rc*cos(t(j))+1;
cm(j)=rc*cos(t(j))-1;
end
plot(c,s,cp,s,cm,s)
%%%%%%%%%%%%%%%%%%%%%%%%%%%%%%%%%%%%%%%%%%%%%%%%%
% extrema
%%%%%%%%%%%%%%%%%%%%%%%%%%%%%%%%%%%%%%%%%%%%%%%%%
tt=linspace(-0.02,0.02,N);
for j=1:N
```

```
hh(j)=2/(3*sqrt(3));
hhh(j)=-2/(3*sqrt(3));
end
plot(tt,hh,tt,hhh)
```

The following Matlab script generates Figure 8.6

```
%%%%%%%%%%%%%%%%%%%%%%%%%%%%%%%%%%%%%%
% p1.m: fixed points u as a function of lambda
%%%%%%%%%%%%%%%%%%%%%%%%%%%%%%%%%%%%%%
clear u* l*
M=1/sqrt(3);
u1=linspace(-1.5,-M,51);
u3=linspace(M,1.5,51);
for j=1:51
lam1(j)=u1(j)^3-u1(j);
lam3(j)=u3(j)^3-u3(j);
end
N2=21;
u2=linspace(-M,M,N2);
for j=1:N2
lam2(j)=u2(j)^3-u2(j);
end
plot(lam1,u1)
hold on
plot(lam2,u2,'o')
hold on
plot(lam3,u3)
text(-0.5, -1, 'u_1^*(\lambda)')
text(0,0.2, 'u_2^*(\lambda)')
text(0.5, 1, 'u_3^*(\lambda)')
xlabel('\lambda')
ylabel('Fixed Points u of du/dt=u(1+u)(1-u)+\lambda')
```

The following Matlab script generates Figure 8.7

```
%%%%%%%%%%%%%%%%%%%%%%%%%%%%%%%%%%%%%%
% p3.m: hysteresis for slowly varying lambda
%%%%%%%%%%%%%%%%%%%%%%%%%%%%%%%%%%%%%%
f=@(t,y) y*(1+y)*(1-y)+sin(2*pi*t/100);
clear u t lam;
t=[0 200];
```

```
u0=1;
[t u]=ode15s(f,t,u0);
J=1000;
tt=linspace(0,200, J);
for j=1:J
lam(j)=sin(2*pi*tt(j)/100);
end
plot(t,u,tt,lam,'-.')
xlabel('time t')
ylabel('\lambda(t) and u(t)')
title('u(t) for slowly varying \lambda')
```

The following Matlab script generates Figure 8.8

```
%%%%%%%%%%%%%%%%%%%%%%%%%%%%%%%%%%%%%%%%%%%%
% p4.m: hysteresis loop for slowly varying lambda
%%%%%%%%%%%%%%%%%%%%%%%%%%%%%%%%%%%%%%%%%%%%
f=@(t,y) y*(1+y)*(1-y)+sin(2*pi*t/100);
clear u t lam;
t=[0 200];
u0=1;
[t u]=ode15s(f,t,u0);
J=length(t);
for j=1:J
lam(j)=sin(2*pi*t(j)/100);
end
plot(lam,u)
xlabel('\lambda(t)')
ylabel('u(t)')
title('Pairs (\lambda(t),u(t)) for parameter dependent ODE')
```

Chapter 10

Two Oscillators: Periodicity, Ergodicity, and Phase Locking

Summary: In this section, we consider two oscillators obeying simple dynamical equations. The state of the two oscillators is described by a pair of angles,

$$\Big(\theta_1(t), \theta_2(t)\Big) .$$

As time progresses one obtains a trajectory on a 2–torus. In the simplest case, the oscillators are uncoupled and obey the dynamical equations

$$\theta_1'(t) = \omega_1, \quad \theta_2'(t) = \omega_2 , \tag{10.1}$$

where $\omega_1 > 0$ and $\omega_2 > 0$ are two fixed frequencies. We will show the following result:

Theorem 10.1. *a) If ω_2/ω_1 is rational, then every orbit $(\theta_1(t), \theta_2(t))$ is periodic.*
b) If ω_2/ω_1 is irrational, then every orbit $(\theta_1(t), \theta_2(t))$ is ergodic.

Here, roughly, ergodicity of the orbit $(\theta_1(t), \theta_2(t))$ means the following: The average time which the orbit spends in any given subset Q of the 2–torus is proportional to the measure of Q. A time average equals a space average. We will explained this in Section 3..

Clearly, under arbitrarily small perturbations of the frequencies ω_j, the quotient ω_2/ω_1 can change from being rational to becoming irrational and vice versa. The theorem then implies that the qualitative behavior of the uncoupled system (10.1) is highly sensitive to perturbations of the frequencies. We treat some coupled oscillators in Section 5.. Coupling often makes the dynamics more robust.

1. The Circle and the Two–Torus

It will be convenient to think of an angle θ as an element of

$$S^1 := \mathbb{R} \bmod 1$$

instead of $\mathbb{R} \bmod 2\pi$. This just saves a lot of writings of 2π. Here we recall that the set $S^1 = \mathbb{R} \bmod 1$ consists of all real numbers where two numbers $x, y \in \mathbb{R}$ are identified if they differ by an integer, i.e., if $x = y + n$ for some $n \in \mathbb{Z}$. Topologically, the space $S^1 = \mathbb{R} \bmod 1$ can be identified with the unit circle.

The set of ordered pairs (θ_1, θ_2) with $\theta_1, \theta_2 \in S^1 = \mathbb{R} \bmod 1$ form the two–torus:

$$S^1 \times S^1 =: T^2 .$$

We can think of T^2 as the unit square $[0,1] \times [0,1]$ where opposite sides are identified, i.e., the point $(1, \theta_2)$ is identified with $(0, \theta_2)$ and the point $(\theta_1, 1)$ is identified with $(\theta_1, 0)$. See Figure 10.1.

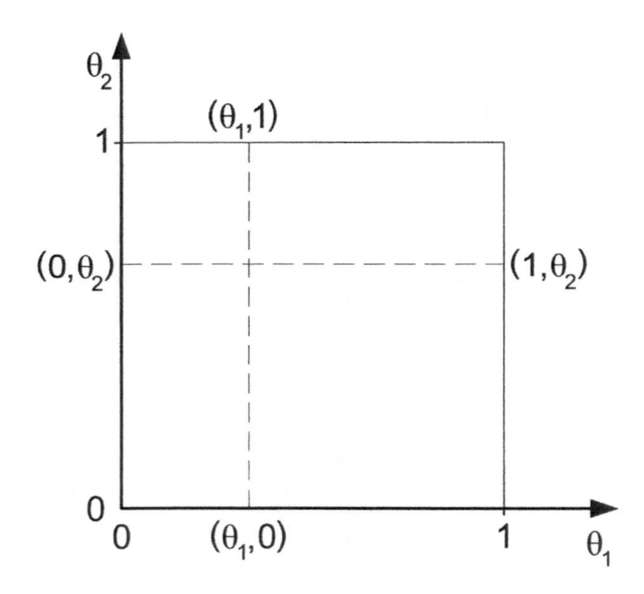

Figure 10.1. The unit square with points to be identified

Example 1: Consider the initial value problem

$$\begin{aligned} \theta_1' &= 2, \quad \theta_1(0) = 0 \\ \theta_2' &= 3, \quad \theta_2(0) = 0 \end{aligned}$$

with solution

$$\theta_1(t) = (2t) \bmod 1, \quad \theta_2(t) = (3t) \bmod 1 \ .$$

With computations modulo 1 we have

$$\begin{aligned}
\Big(\theta_1(1/3), \theta_2(1/3)\Big) &= (2/3, 1) = (2/3, 0) =: P \\
\Big(\theta_1(1/2), \theta_2(1/2)\Big) &= (1, 1/2) = (0, 1/2) =: Q \\
\Big(\theta_1(2/3), \theta_2(2/3)\Big) &= (1/3, 1) = (1/3, 0) =: R \\
\Big(\theta_1(1), \theta_2(1)\Big) &= (2, 3) = (0, 0)
\end{aligned}$$

It is clear that the solution has the period $T = 1$.

Figure 10.2 shows the trajectory on the unit square. If one identifies the points $(0, \theta_2)$ and $(1, \theta_2)$, the square becomes a cylinder. In Figure 10.3 we show the same trajectory, but now on the cylinder. If we further identify the bottom of the cylinder with its top, we obtain the 2–torus $T^2 = S^1 \times S^1$. The trajectory then moves on this 2–torus.

2. Uncoupled Oscillators: Periodic Solutions

Consider the system

$$\theta_1' = \omega_1, \quad \theta_2' = \omega_2 \tag{10.2}$$

where $\omega_1 > 0$ and $\omega_2 > 0$ are two constant frequencies. If we give an initial condition

$$\theta_1(0) = \theta_{10}, \quad \theta_2(0) = \theta_{20}$$

then the solution is

$$\begin{aligned}
\theta_1(t) &= (\theta_{10} + \omega_1 t) \bmod 1 \\
\theta_2(t) &= (\theta_{20} + \omega_2 t) \bmod 1
\end{aligned}$$

Assume that the solution has period $T > 0$. We then have, with positive integers m and n:

$$\theta_{10} + \omega_1 T = \theta_{10} + m, \quad \theta_{20} + \omega_2 T = \theta_{20} + n \ ,$$

which implies that

$$\omega_1 T = m, \quad \omega_2 T = n \ .$$

Therefore, if the system (10.2) has a periodic solution, then the ratio

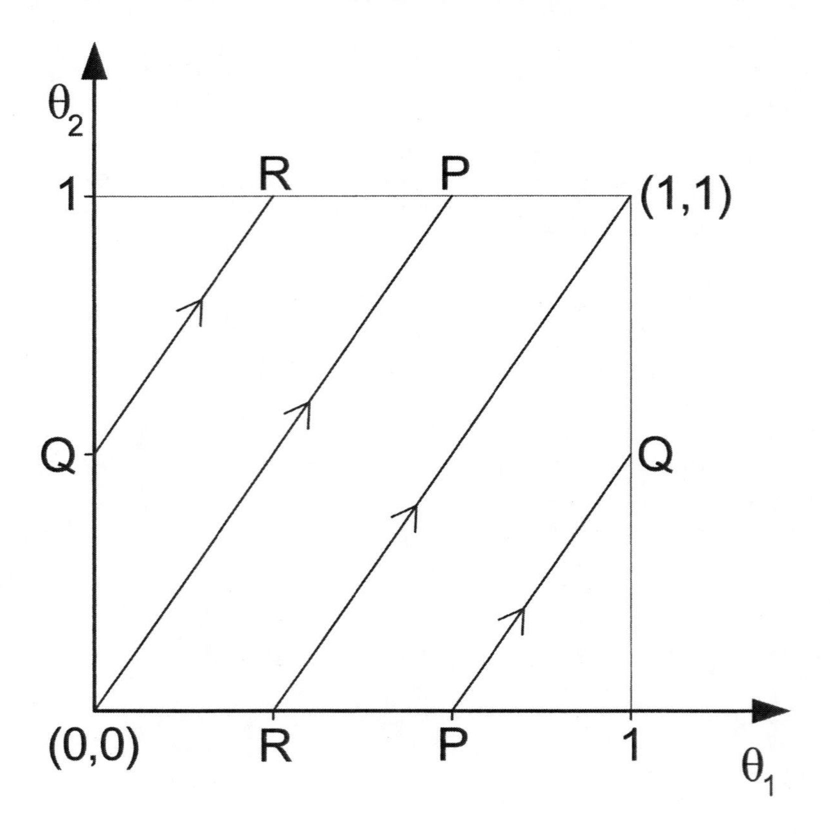

Figure 10.2. The trajectory on the unit square

$$\frac{\omega_1}{\omega_2} = \frac{m}{n} \quad \text{with} \quad m, n \in \mathbb{N} \tag{10.3}$$

is a rational number.

Conversely, assume that ω_1/ω_2 is rational, i.e., (10.3) holds. Consider the solution of $\theta_1' = \omega_1, \theta_2' = \omega_2$ with initial data $\theta_1(0) = \theta_{10}, \theta_2(0) = \theta_{20}$ at time $T = m/\omega_1 = n/\omega_2$. We have

$$\begin{aligned}
\theta_1(T) &= \theta_{10} + \omega_1 T = \theta_{10} + m = \theta_{10} \bmod 1 \\
\theta_2(T) &= \theta_{20} + \omega_2 T = \theta_{20} + n = \theta_{20} \bmod 1
\end{aligned}$$

thus

$$\theta_1(T) = \theta_1(0) \quad \text{and} \quad \theta_2(T) = \theta_2(0) \ .$$

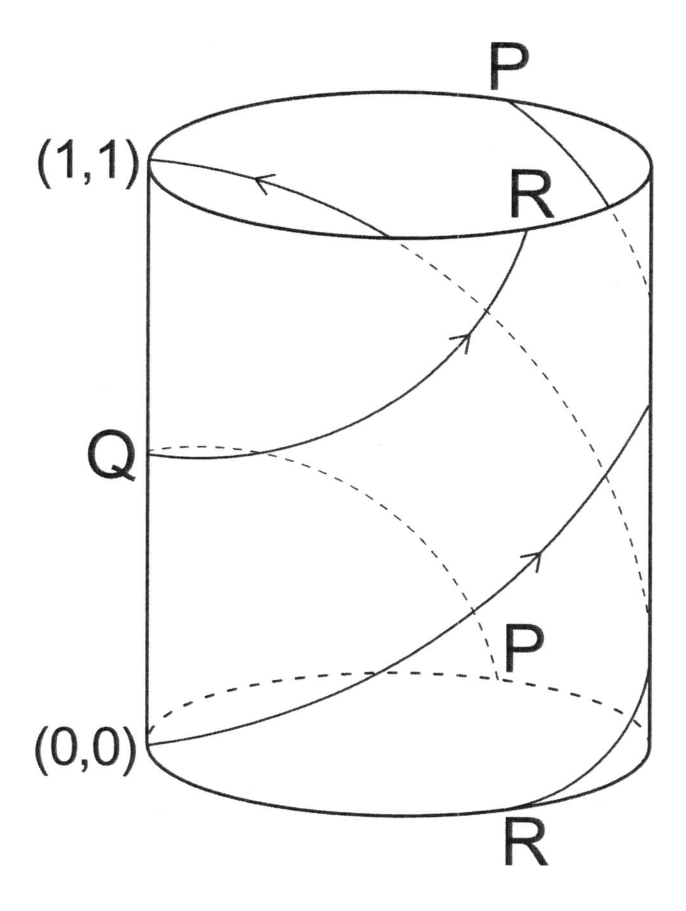

Figure 10.3. The trajectory on the unit square

The solution has period $T = m/\omega_1 = n/\omega_2$.

We summarize our result:

Lemma 10.1. *Consider the system (10.2) where $\omega_1 > 0$ and $\omega_2 > 0$ are constant frequencies. The following conditions are equivalent:*

1. *There is a periodic solution.*
2. *The ratio $\frac{\omega_2}{\omega_1}$ is rational.*
3. *Every solution is periodic.*

3. Uncoupled Oscillators: Ergodicity

Let us consider the system

$$\theta_1' = \omega_1, \quad \theta_2' = \omega_2 \tag{10.4}$$

where $\omega_1 > 0$ and $\omega_2 > 0$ are constant frequencies and assume now that

$$\frac{\omega_1}{\omega_2} \in \mathbb{R} \setminus \mathbb{Q} ,$$

i.e., the ratio of the frequencies is irrational. For any initial condition

$$\theta_1(0) = \theta_{10}, \quad \theta_2(0) = \theta_{20}$$

the solution formula

$$\theta_1(t) = (\theta_{10} + \omega_1 t) \bmod 1, \quad \theta_2(t) = (\theta_{20} + \omega_2 t) \bmod 1 ,$$

holds.

Let us denote the solution of the system simply by

$$\theta(t) = \Big(\theta_1(t), \theta_2(t)\Big) . \tag{10.5}$$

If we fix any subset Q of the torus T^2 we can ask: How much time, on average, does the solution $\theta(t)$ spend in the set Q? To get a precise definition, let us first fix a time $T > 0$ and set [1]

$$M(Q,T) = \{t \ : \ 0 \le t \le T, \ \theta(t) \in Q\} ,$$

i.e., $M(Q,T)$ is the set of all times t in the interval $0 \le t \le T$ for which $\theta(t)$ lies in Q.

We then consider the limit [2]

$$\lim_{T\to\infty} \frac{1}{T} \, measure(M(Q,T)) . \tag{10.6}$$

If the limit (10.6) exists and equals $L = 0.25$, for example, then, in the long run, the solution will spend (about) a quarter of its time in the set Q.

We will prove the following result:

[1] The set of times $M(Q,T)$ also depends on the solution $\theta(t)$ under consideration, but this is suppressed in our notation.

[2] Readers familiar with measure theory may think of the Lebesgue measure in formula (10.6). However, we will restrict our considerations to the case where Q is a rectangle. The set of times $M(Q,T)$ then is a finite union of subintervals of the real line and the measure of $M(Q,T)$ is simply the sum of the lengths of these intervals.

Theorem 10.2. *Let $Q = [a_1, b_1] \times [a_2, b_2] \subset T^2$ denote a rectangle. If $\frac{\omega_1}{\omega_2}$ is irrational, then the limit (10.6) exists and*

$$\lim_{T\to\infty} \frac{1}{T}\, measure(M(Q,T)) = area(Q)\ . \tag{10.7}$$

In the next section, we will prove, in detail, a one–dimensional version of the above theorem. Theorem 10.2 can be proved by extending the arguments to two dimensions.

Remarks on Ergodicity. Theorem 10.2 connects a time average (10.6) with a space average, namely the size of the set Q, which is a subset of the state space T^2. Using measure theory, one can extend the formula (10.7) from rectangles Q to more general measurable subsets Q of T^2. Theorem 10.2 is an example of *Ergodic Theory*, which applies measure theoretic ideas to dynamics.

A historical starting point of ergodic theory was the work of Ludwig Boltzmann (1844-1906). He began the challenging task to derive the laws of thermodynamics from the laws of mechanics and modeled a gas as a collection of a large number of bouncing balls, which we now call molecules. Since the number N of molecules in any macro volume of a gas is very large ($N \sim 10^{23}$), Boltzmann applied probabilistic arguments.

Boltzmann's famous formula

$$S = k \log W$$

connects the entropy S of a macrostate and the probability W of the state. The constant $k = 1.3806505 * 10^{-23} J/K$ is Boltzmann's constant.

Boltzmann *assumed* that time averages can be computed as volumes in phase–space, an assumption that became known as Boltzmann's Ergodic Hypothesis. For Boltzmann's system, the validity of this hypothesis has never been justified, however. The systems (10.4) with irrational ratio ω_1/ω_2 are ergodic. This is the essence of Theorem 10.2.

4. Time Average Equals Space Average for a Circle Map

4.1. A Poincaré Map

Let us first introduce the Poincaré map [3] for a system

$$\theta_1' = \omega_1, \quad \theta_2' = \omega_2 \tag{10.8}$$

which we considered in the two previous sections. Recall that $\omega_1 > 0$ and $\omega_2 > 0$ are fixed frequencies. If we give an initial condition

$$\theta_1(0) = 0, \quad \theta_2(0) = \beta$$

[3] named after the French mathematician Henri Poincaré (1854–1912)

then the solution is

$$\theta_1(t) = (\omega_1 t) \bmod 1, \quad \theta_2(t) = (\beta + \omega_2 t) \bmod 1 .$$

At time $T = 1/\omega_1$ we have

$$\theta_1(T) = 1 \bmod 1 = 0 = \theta_1(0) ,$$

i.e., at time $T = 1/\omega_1$ the angle θ_1 has returned to its value at time $t = 0$. At the same time $T = 1/\omega_2$, the angle θ_2 has changed from β to

$$\theta_2(T) = \left(\beta + \frac{\omega_2}{\omega_1}\right) \bmod 1 .$$

The circle map

$$P : \begin{cases} S^1 & \rightarrow \quad S^1 \\ \beta & \rightarrow \quad \left(\beta + \frac{\omega_2}{\omega_1}\right) \bmod 1 \end{cases} \tag{10.9}$$

is called the Poincaré map of the system (10.8). In Figure 10.4 we show the pairs of angles $(0, \beta)$ and $(1, P(\beta))$, where the assignment $\beta \rightarrow P(\beta)$ is the Poincaré map corresponding to the system (10.8).

To study a Poincaré map of a system is often easier than to study the system itself because the Poincaré map works in a lower dimension. In the present case, the solution of (10.8) moves on the two–dimensional torus $S^1 \times S^1 = T^2$, but the Poincaré map 10.9 maps the one–dimensional circle S^1 onto itself. That's why we can apply one–dimensional Fourier series in the next section.

4.2. Ergodicity of a Circle Map

In this section we consider a general circle map of the form (10.9). To this end, fix a real number c and define the circle map $\phi = \phi_c$ by

$$\phi : \begin{cases} S^1 & \rightarrow \quad S^1 \\ \beta & \rightarrow \quad (\beta + c) \bmod 1 \end{cases} \tag{10.10}$$

Any seed $\beta_0 \in S^1$ determines the orbit

$$\beta_n = \phi^n(\beta_0), \quad n = 0, 1, 2, \ldots$$

If $I \subset [0, 1)$ is a subinterval of the state space S^1 and if N denotes any positive integer, we define

$$M(I, N) = \{n \in \mathbb{N} \ : \ 0 \leq n \leq N, \ \phi^n(\beta_0) \in I\} .$$

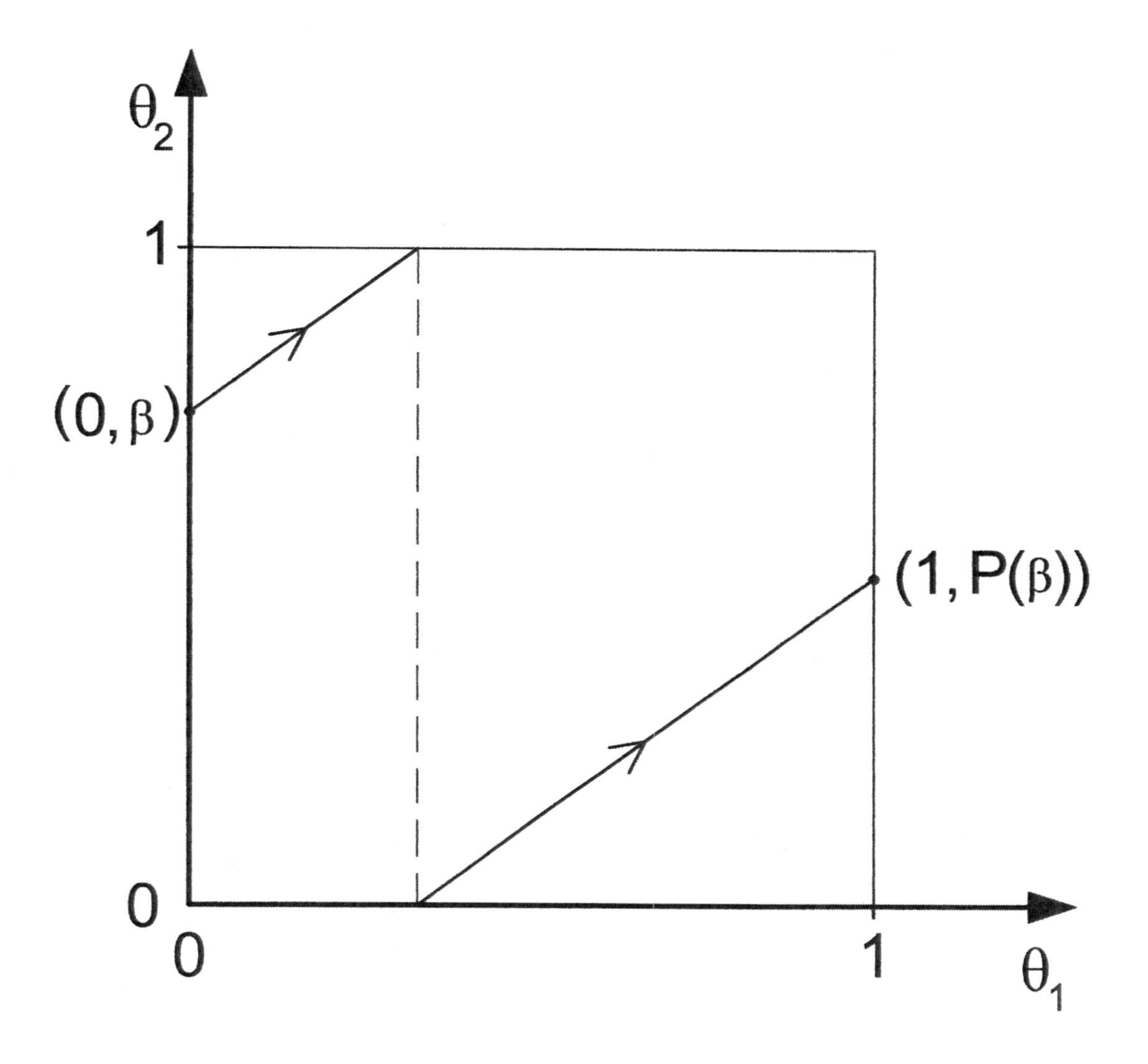

Figure 10.4. The assignment $\beta \to P(\beta)$ is the Poincaré map

The number of elements of the set $M(I, N)$ is denoted by $\#M(I, N)$. For large N, the quotient

$$\frac{1}{N+1} \#M(I, N)$$

is the average proportion of times at which the orbit

$$\beta_n = \phi^n(\beta_0)$$

falls into the interval I. We now state the one–dimensional analogue of Theorem 10.2. This result and its beautiful proof go back to Hermann Weyl [18].

Theorem 10.3. *Consider the circle map (10.10) where c is irrational. If $\beta_n = \phi^n(\beta_0)$ is any orbit and if $I \subset [0,1)$ is any subinterval of the state space, then we have*

$$\lim_{N\to\infty} \frac{1}{N+1} \#M(I,N) = length(I) \; . \tag{10.11}$$

Outline of Proof: First let $f : S^1 \to \mathbb{C}$ denote an arbitrary function. Then, if the following limit exists,

$$\lim_{N\to\infty} \frac{1}{N+1} \sum_{n=0}^{N} f(\phi^n(\beta_0)) \tag{10.12}$$

it gives the long time average value of the function f along the orbit $\phi^n(\beta_0)$. Using this notation and denoting the characteristic function of the interval I by $\chi_I(\theta)$, we can reformulate equation (10.11) as

$$\lim_{N\to\infty} \frac{1}{N+1} \sum_{n=0}^{N} \chi_I(\phi^n(\beta_0)) = \int_0^1 \chi_I(\theta)\, d\theta \; . \tag{10.13}$$

Let us now outline how to prove this formula. First, we will consider Fourier modes

$$f_k(\theta) = e^{2\pi i k\theta}, \quad k \in \mathbb{Z} \; ,$$

and prove the corresponding equation

$$\lim_{N\to\infty} \frac{1}{N+1} \sum_{n=0}^{N} f_k(\phi^n(\beta_0)) = \int_0^1 f_k(\theta)\, d\theta \; . \tag{10.14}$$

Here, as we will see, the irrationality of c is crucial. Using Fourier approximation, it then follows that the formula

$$\lim_{N\to\infty} \frac{1}{N+1} \sum_{n=0}^{N} f(\phi^n(\beta_0)) = \int_0^1 f(\theta)\, d\theta \tag{10.15}$$

holds for any continuous function $f : S^1 \to \mathbb{C}$. The characteristic function χ_I is discontinuous, but an approximation argument will allow us to deduce the desired equation (10.13) from (10.15) for continuous f.

Fourier Modes. We now prove formula (10.14) for the Fourier modes $f_k(\theta)$.

Lemma 10.2. *Assume that c is irrational. The formula (10.14) holds for any Fourier mode $f_k(\theta) = e^{2\pi ik\theta}, k \in \mathbb{Z}$.*

Proof: If $k = 0$ then $f_k = f_0 \equiv 1$. The left–hand side and the right–hand side of (10.14) are both equal to 1.

Now let k denote any integer, but $k \neq 0$. In this case we have

$$\int_0^1 f_k(\theta)\, d\theta = \int_0^1 e^{2\pi ik\theta}\, d\theta = 0 \; ,$$

i.e., the right–hand side of equation (10.14) equals zero. To compute the left–hand side we use the geometric sum formula,

$$1 + q + \ldots + q^N = \frac{q^{N+1} - 1}{q - 1} \quad \text{for} \quad q \neq 1 \; .$$

With

$$q = e^{2\pi ikc}$$

we have

$$\begin{aligned} \sum_{n=0}^{N} f_k(\phi^n(\beta_0)) &= \sum_{n=0}^{N} e^{2\pi ik(\beta_0 + nc)} \\ &= e^{2\pi ik\beta_0} \sum_{n=0}^{N} q^n \\ &= e^{2\pi ik\beta_0} \frac{q^{N+1} - 1}{q - 1} \end{aligned}$$

Note that the irrationality of c ensures that $q \neq 1$.

Since $|q| = 1$ it follows that

$$\Big| \sum_{n=0}^{N} f_k(\phi^n(\beta_0)) \Big| \leq \frac{2}{|q - 1|} \; ,$$

i.e., we have a bound of the sum which is independent of N. Clearly, the left–hand side of (10.14) equals zero for $k \neq 0$. The lemma is proved. ◇

Approximation by Trigonometric Polynomials. Any finite linear combination of Fourier modes $f_k(\theta) = e^{2\pi ik\theta}, k \in \mathbb{Z}$, is called a trigonometric polynomial (of period one). Thus, a trigonometric polynomial takes the general form

$$p_K(\theta) = \sum_{k=-K}^{K} \alpha_k e^{2\pi i k\theta}$$

where K is finite. We denote the linear space of all trigonometric polynomials by $\mathcal{T}$. Since the sum in a trigonometric polynomial is *finite*, it is easy to extend the previous lemma from Fourier modes to trigonometric polynomials:

Lemma 10.3. *Assume that c is irrational and let $p_K \in \mathcal{T}$ denote a trigonometric polynomial. Then we have*

$$\lim_{N\to\infty} \frac{1}{N+1} \sum_{n=0}^{N} p_K(\phi^n(\beta_0)) = \int_0^1 p_K(\theta)\, d\theta \ . \tag{10.16}$$

We now state an approximation theorem of Weierstrass:

Theorem 10.4. *Let $f : \mathbb{R} \to \mathbb{C}$ denote a continuous, 1–periodic function. There exists a sequence $p_K \in \mathcal{T}$ with*

$$\lim_{K\to\infty} |f - p_K|_\infty = 0 \ . \tag{10.17}$$

Here we have used the notation

$$|g|_\infty = \max_\theta |g(\theta)|$$

for the maximum norm of a function.

It is easy to show that (10.17) implies

$$\int_0^1 p_K(\theta)\, d\theta \to \int_0^1 f(\theta)\, d\theta \quad \text{as} \quad K \to \infty \ .$$

One then may try to prove the formula (10.15) by letting $K \to \infty$ in (10.16). This is the right idea, but requires to exchange the limits $N \to \infty$ and $K \to \infty$.

Instead of justifying this exchange of limits, we will proceed somewhat differently.

Rigorous Arguments. Recall that $\beta_n = \phi^n(\beta_0)$ denotes an orbit of the circle map (10.10) where c is an irrational number. For any $f : \mathbb{R} \to \mathbb{C}$ which is 1–periodic, let us denote

$$A_N f = \frac{1}{N+1} \sum_{n=0}^{N} f(\phi^n(\beta_0)) \ , \tag{10.18}$$

and let

$$Af = \lim_{N\to\infty} A_N f$$

if the limit exists. We note the trivial estimates

$$|A_N f| \le |f|_\infty \tag{10.19}$$

and

$$|Af| \le |f|_\infty \; . \tag{10.20}$$

With these notations, we now reformulate equation (10.15):

Lemma 10.4. *Assume that c is irrational, and let $f : \mathbb{R} \to \mathbb{C}$ denote a continuous, 1–periodic function. Then we have*

$$Af = \int_0^1 f(\theta)\, d\theta \; . \tag{10.21}$$

Proof: We first show that $A_N f$ is a Cauchy sequence and let $\varepsilon > 0$ be given. By Theorem 10.4 there exists $p \in \mathcal{T}$ with $|f - p|_\infty < \varepsilon$. We have

$$\begin{aligned} |A_N f - A_M f| &\le |A_N f - A_N p| + |A_N p - A_M p| + |A_M p - A_M f| \\ &\le 2\varepsilon + |A_N p - A_M p| \\ &\le 3\varepsilon \end{aligned}$$

for $N, M \ge N(\varepsilon)$. We have used the estimate (10.19) with f replaced by $f - p$. The last estimate, $|A_N p - A_M p| < \varepsilon$ for large N, M, follows from Lemma 10.3. These arguments shows that $\lim A_N g = Ag$ exists for any continuous, 1–periodic function $g : \mathbb{R} \to \mathbb{C}$.

We now apply Weierstrass Theorem 10.4 and choose a sequence $p_K \in \mathcal{T}$ with $|f - p_K|_\infty \to 0$. We have

$$A_N f = A_N(f - p_K) + A_N p_K \; .$$

In this equation let $N \to \infty$ to obtain

$$Af = A(f - p_K) + \int_0^1 p_K\, dx \; .$$

Here $|A(f - p_K)| \le |f - p_K|_\infty \to 0$. In the above equation, we let $K \to \infty$. This finally yields

$$Af = \int_0^1 f(x)\, dx \; .$$

The lemma is proved. $\diamond$.

Approximation of the Characteristic Function. Recall that $I \subset [0,1)$ denotes a subinterval of the state space $S^1 = \mathbb{R} \bmod 1$, which we can identify with the half–open interval $[0,1)$. Lemma 10.4 does not apply to the characteristic function $\chi_I(\theta)$, which is discontinuous. To prove Theorem 10.3 we approximate $f(\theta) := \chi_I(\theta)$ by piecewise linear, continuous functions $f_\varepsilon(\theta)$ and $g_\varepsilon(\theta)$ from above and from below.

For example, if $I = [a,b] \subset (0,1)$, then we choose $f_\varepsilon(\theta)$ as the function which is zero outside the interval $[a-\varepsilon, b+\varepsilon]$, which is one for $a \le \theta \le b$, and which is linear for $a-\varepsilon \le \theta \le a$ and for $b \le \theta \le b+\varepsilon$. We then have

$$g_\varepsilon(\theta) \le f(\theta) \le f_\varepsilon(\theta) \quad \text{for all} \quad \theta$$

and

$$\int_0^1 g_\varepsilon(\theta)\, d\theta = l - \varepsilon \quad \text{and} \quad \int_0^1 f_\varepsilon(\theta)\, dx = l + \varepsilon \quad \text{with} \quad l = \text{length}(I) \; .$$

With the averaging operator A_N defined in (10.18) we have

$$A_N g_\varepsilon \le A_N f \le A_N f_\varepsilon \; ,$$

which implies that

$$|A_N f - A_N f_\varepsilon| \le |A_N(f_\varepsilon - g_\varepsilon)| \; .$$

We now show that $A_N f$ is a Cauchy sequence. Note that

$$\begin{aligned} |A_N f - A_M f| &\le |A_N f - A_N f_\varepsilon| + |A_N f_\varepsilon - A_M f_\varepsilon| + |A_M f_\varepsilon - A_M f| \\ &\le |A_N(f_\varepsilon - g_\varepsilon)| + |A_N f_\varepsilon - A_M f_\varepsilon| + |A_M(f_\varepsilon - g_\varepsilon)| \end{aligned}$$

Here

$$A_N(f_\varepsilon - g_\varepsilon) \to 2\varepsilon \quad \text{as} \quad N \to \infty$$

and

$$|A_N f_\varepsilon - A_M f_\varepsilon| < \varepsilon$$

for $N, M \ge N(\varepsilon)$. Therefore,

$$|A_N f - A_M f| \le 5\varepsilon \quad \text{if} \quad N, M \ge N(\varepsilon) \; ,$$

showing that $A_N f$ is a Cauchy sequence.

Recall the bounds

$$A_N g_\varepsilon \le A_N f \le A_N f_\varepsilon$$

and let $N \to \infty$ to obtain

$$l - \varepsilon \le Af \le l + \varepsilon \ .$$

Since $\varepsilon > 0$ was arbitrary, we have shown that

$$Af = l = length(I) \ .$$

This completes a rigorous proof of Theorem 10.3. $\diamond$

Extension to Two Dimensions. To prove Theorem 10.2 one can proceed in the same way using the two–dimensional Fourier modes

$$f_k(\theta) = e^{2\pi i(k_1\theta_1 + k_2\theta_2)}, \quad \theta = (\theta_1, \theta_2) \in T^2 \ .$$

The main point is the following: If ω_1/ω_2 is irrational, then

$$k_1\omega_1 + k_2\omega_2 \neq 0$$

for

$$k = (k_1, k_2) \in \mathbb{Z}^2, \quad k \neq (0,0) \ .$$

One then obtains that

$$\lim_{T\to\infty} \frac{1}{T} \int_0^T f_k(\theta_{10} + \omega_1 t, \theta_{20} + \omega_2 t)\, dt = \begin{cases} 1 & \text{for } k = (0,0) \\ 0 & \text{for } k \neq (0,0) \end{cases}$$

This extends Lemma 10.2 to two dimensions. The remaining arguments can also be carried over from one to two dimension.

5. Coupled Oscillators

Consider the system (see Section 8.6 of [16]):

$$\theta_1' = \omega_1 + K_1 \sin(\theta_2 - \theta_1) \quad (10.22)$$

$$\theta_2' = \omega_2 + K_2 \sin(\theta_1 - \theta_2) \quad (10.23)$$

In this section, we consider the variables θ_j as elements of $\mathbb{R} \bmod 2\pi$ since the function $\sin\theta$ has period 2π.

For $K_1 = K_2 = 0$ we obtain the uncoupled system (10.1), but for $K_j \neq 0$ the two oscillators are coupled.

For definiteness, assume that

$$\omega_1 > \omega_2 > 0, \quad K := K_1 + K_2 > 0 \ .$$

Our aim here is to show that a sufficiently strong coupling eliminates the delicate dependency of the dynamics on ω_1/ω_2, which was stated in Theorem 10.1.

Let

$$\phi = \theta_1 - \theta_2$$

denote the phase difference. We obtain

$$\phi' = \omega_1 - \omega_2 - K \sin\phi =: f(\phi) \ .$$

Let us now assume that the coupling is so strong that

$$K > \omega_1 - \omega_2 > 0 \ .$$

Then the fixed point equation $f(\phi) = 0$ has two solutions $\phi^*_{1,2}$ with

$$0 < \phi^*_1 < \frac{\pi}{2} < \phi^*_2 < \pi \ .$$

The fixed point ϕ^*_1 is asymptotically stable whereas ϕ^*_2 is unstable. If we disregard the exceptional initial condition $\phi(0) = \phi^*_2$, then

$$\phi(t) \to \phi^*_1 \quad \text{as} \quad t \to \infty \ .$$

Therefore, except for an initial transient, the given coupled system (10.22), (10.23) is well approximated by

$$\theta'_1 = \omega_1 - K_1 \sin\phi^*_1 \tag{10.24}$$

$$\theta'_2 = \omega_2 + K_2 \sin\phi^*_1 \tag{10.25}$$

Using the equation

$$\sin\phi^*_1 = \frac{1}{K}(\omega_1 - \omega_2)$$

one finds that

$$\omega_1 - K_1 \sin\phi^*_1 = \frac{1}{K}(K_1\omega_2 + K_2\omega_1) =: \omega^*$$

and, similarly,

$$\omega_2 + K_2 \sin\phi^*_1 = \omega^* \ .$$

The system (10.24), (10.25) thus reads

$$\begin{aligned} \theta_1' &= \omega^* \\ \theta_2' &= \omega^* \end{aligned}$$

One obtains that, after an initial transient, the oscillators move (approximately) with the same frequency ω^*. The phase difference $\theta_1(t) - \theta_2(t)$ approaches a constant,

$$\theta_1(t) - \theta_2(t) = \phi(t) \to \phi_1^* \quad \text{as} \quad t \to \infty \; .$$

Except for one unstable trajectory, all other trajectories approach periodic motion with frequency ω^* and phase difference ϕ_1^*. One calls this a *phase locked motion.*

Chapter 11

The Gambler's Ruin Problem

Summary: The Gambler's Ruin Problem is an example of a discrete–time Markov chain. After describing the game and looking at a numerical realization (using Matlab's random number generator) we will introduce the so–called transition matrix P of the game. The matrix P encodes the probabilities of the random evolution step $X_t \to X_{t+1}$. Interestingly, though the evolution is random, the probability density vectors $\pi^{(t)}$ evolve deterministically:

$$\pi^{(t+1)} = P\pi^{(t)} \ .$$

We will use this to answer some simple questions about the game.

1. Description of the Game

We take here a naive approach to the concepts of randomness and probability and assume that we can flip a coin which shows heads (H) or tails (T) *randomly.* Here H comes up with probability p, where $0 < p < 1$, and T comes up with probability $q = 1 - p$. The coin is unfair unless $p = \frac{1}{2}$. The outcome of coin tosses will determine how the game proceeds.

Fix some integer N, for example $N = 1,000$, and let the set

$$A = \{0, 1, \ldots, N\}$$

denote the state space. In our game, the time axis is discrete:

$$\mathcal{T} = \{0, 1, 2, \ldots\} \ .$$

At time $t = 0$ the gambler starts with capital $X_0 = k \in A$, and we now describe how his capital $X_t \in A$ evolves as time progresses. The variable X_t denotes the gambler's capital at time t.

There are three simple rules:

Rule 1: If $X_t = 0$ then the gambler is ruined at time t. Therefore, $X_{t+1} = 0$.

Rule 2: If $X_t = N$ then the gambler goes home and $X_{t+1} = N$.

Rule 3: If $X_t \in \{1, 2, \dots, N-1\}$ then the gambler loses a dollar if the coin toss shows T (with probability q) and wins a dollar if H shows up (with probability p). Therefore, if $X_t \in \{1, 2, \dots, N-1\}$, then

$$X_{t+1} = \begin{cases} X_t + 1 \text{ with probability } p \\ X_t - 1 \text{ with probability } q \end{cases} .$$

The game is specified by the number N, which is the maximal capital that can be reached, the start capital $X_0 = k$, and the probabilities p and q with $p + q = 1$.

The capital X_t is an example of a random variable taking values in the finite state space $A = \{0, 1, \dots, N\}$.

2. Some Questions and Numerical Realization

One can ask many questions. For example, let

$$p = 0.49, \quad q = 0.51 ,$$

and let

$$k = 800, \quad N = 1,000 .$$

What is the probability for the gambler to reach $X_t = N = 1,000$ and to go home having won 200 Dollars? What is the probability to first get ruined, i.e., to reach $X_t = 0$ and to go home having lost 800 Dollars? How long, on average, does the game last before $X_t = N$ or $X_t = 0$ is reached?

In the next chapter, we will learn how to answer these questions by analysis. However, if the game or the questions become more difficult, it is often a good idea to run many realizations of the game. Of course, it is then tedious to actually make coin tosses. Matlab's random number generator is a great help. The command $Y = rand$ gives a random number Y in the interval $0 \le Y \le 1$ and the Ys are uniformly distributed in this interval. Therefore, we have $0 \le Y \le p$ with probability p and $p < Y \le 1$ with probability $q = 1 - p$. In Figure 11.1 we show one realization of the game with $N = 20, k = 10, p = q = \frac{1}{2}$ computed with the following code.

```
%%%%%%%%%%%%%%%%%%%%%%%%%%%%%%%%%%%%%%%%%%%%%%%%%%%%%%%%%%%%%
%g3.m realisation of gambler's ruin
%%%%%%%%%%%%%%%%%%%%%%%%%%%%%%%%%%%%%%%%%%%%%%%%%%%%%%%%%%%%%
clear X t bottom top
% specification of parameters
```

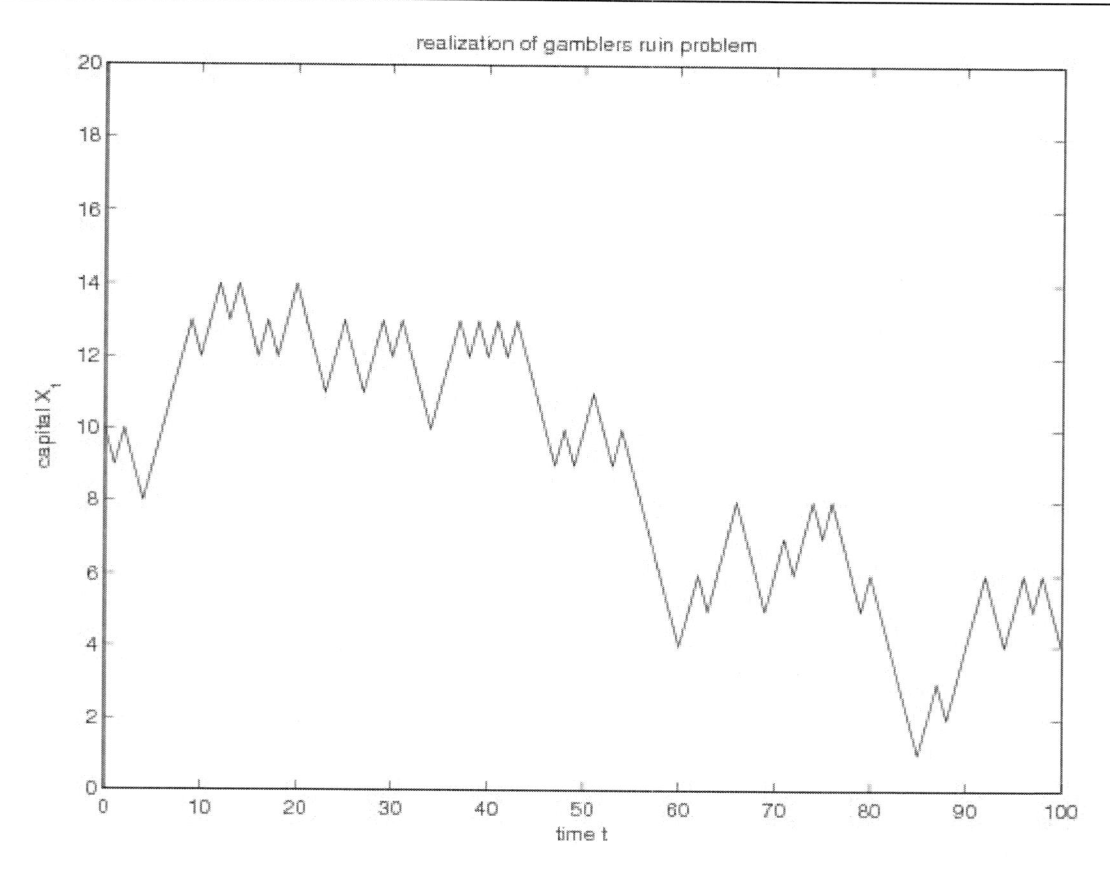

Figure 11.1. Realization of random evolution in gambler's ruin problem

```
p=0.5;
N=20;
k=10;
nmax=100;
%%%%%%%%%%%%%%%%%%%%%%%%%%%%%%%%%%%%%%%%%%%%%%%%%%%%%%%%%%%%
X(1)=k;
for j=2:(nmax+1)
g=X(j-1);
if (g==0)||(g==N)
X(j)=g;
else
y=rand;
if (y<p)
```

```
X(j)=g+1;
else
X(j)=g-1;
end
end
end
%%%%%%%%%%%%%%%%%%%%%%%%%%%%%%%%%%%%%%%%%%%%%%%%%%%%%%%%%%%%%%
for j=1:nmax+1
t(j)=j-1;
bottom(j)=0;
top(j)=N;
end
%%%%%%%%%%%%%%%%%%%%%%%%%%%%%%%%%%%%%%%%%%%%%%%%%%%%%%%%%%%%%%
plot(t,X,t,bottom,t,top)
xlabel('time t')
ylabel('capital X_t')
title('realization of gamblers ruin problem')
```

3. The Transition Matrix P

Let us develop some mathematical tools to analyze random evolutions taking the gambler's ruin game as an example. First, we want to encode the three rules (see Section 1.) that describe the evolution step

$$X_t \to X_{t+1}$$

in a so–called transition matrix:

$$P = (p_{ij})_{0 \le i,j \le N} \in \mathbb{R}^{(N+1)\times(N+1)} \; .$$

Here the matrix P has size $(N+1)\times(N+1)$ because the state space A has $N+1$ elements. By definition, the matrix element p_{ij} is

$$p_{ij} = prob\Big(X_{t+1} = i \;\Big|\; X_t = j\Big) \; .$$

In words: p_{ij} is the probability that $X_{t+1} = i$ under the assumption that $X_t = j$. In our game, the rules for the transition $X_t \to X_{t+1}$ do not depend explicitly on time t; therefore, the matrix P is also independent of t.

What does P look like for our game? First let $j = 0$, i.e., assume that $X_t = 0$. Rule 1 says that $X_{t+1} = 0$. Therefore,

$$p_{i0} = \left\{ \begin{array}{ll} 1 & \text{for } i = 0 \; , \\ 0 & \text{for } 1 \le i \le N \; . \end{array} \right.$$

This gives us the 0–th column of P:

$$(p_{i0})_{i=0,\dots,N} = \begin{pmatrix} 1 \\ 0 \\ \vdots \\ 0 \end{pmatrix} .$$

In the same way, Rule 2 gives us the last column of P,

$$(p_{iN})_{i=0,\dots,N} = \begin{pmatrix} 0 \\ \vdots \\ 0 \\ 1 \end{pmatrix} ,$$

and Rule 3 gives us the j–th column of P for $1 \le j \le N-1$. One obtains:

$$P = \begin{pmatrix} 1 & q & 0 & \dots & 0 \\ 0 & 0 & q & & 0 \\ \vdots & p & 0 & q & \vdots \\ \vdots & & \ddots & \ddots & 0 \\ 0 & \dots & 0 & p & 1 \end{pmatrix} . \tag{11.1}$$

General Remarks on Transition Matrices: Here is an important concept for transition matrices:

Definition 11.1: A real square matrix $P = (p_{ij})$ is called *column stochastic* if $p_{ij} \ge 0$ for all i, j and if all column sums of P equal one.

Since $0 < p < 1$ and $p + q = 1$ it is clear that the matrix (11.1) is column stochastic. To apply matrix algebra, we introduce the vector

$$\mathbf{e} = \begin{pmatrix} 1 \\ 1 \\ \vdots \\ 1 \end{pmatrix} .$$

Then, if P is column stochastic, we have

$$P^T \mathbf{e} = \mathbf{e} ,$$

where P^T denotes the transpose of P. The vector $\mathbf{e}$ is an eigenvector of P^T for the eigenvalue $\lambda_1 = 1$. As we know from linear algebra, the eigenvalues λ_j of P^T are also eigenvalues of P. Thus, if P is column stochastic, then $\lambda_1 = 1$ is an eigenvalue of P. Thus there is a non–zero vector π with $P\pi = \pi$. If π is unique (after normalization) then, as we will see, this vector π is very important: It gives us the stationary probability density of the random process. We will illustrate this by an example in Chapter 15. For the matrices (11.1) the eigenvalue $\lambda_1 = 1$ is geometrically and algebraically double, however. This will be exploited this in the next chapter.

4. Evolution of Probability Density

In the section we will give an important application of the transition matrix P. We use the notation

$$\mathbf{e}_0, \mathbf{e}_1, \ldots, \mathbf{e}_N$$

for the standard basis of column vectors of $\mathbb{R}^{N+1}$. For example,

$$\mathbf{e}_0 = \begin{pmatrix} 1 \\ 0 \\ 0 \\ \vdots \\ 0 \end{pmatrix}, \quad \mathbf{e}_1 = \begin{pmatrix} 0 \\ 1 \\ 0 \\ \vdots \\ 0 \end{pmatrix}.$$

Recall that our game depends on the parameters N (the maximal capital), k (the starting capital), and the probabilities p and q with $p + q = 1$. Think of these parameters as being fixed. We then introduce the probability density vectors

$$\pi^{(t)} \in \mathbb{R}^{N+1}, \quad t = 0, 1, 2, \ldots$$

for the random evolution $X_t \to X_{t+1}$. By definition,

$$\pi_j^{(t)} = prob(X_t = j), \quad 0 \le j \le N ,$$

i.e., $\pi_j^{(t)}$ is the probability that, at time t, the capital X_t equals j. For example, we have

$$\pi^{(0)} = \mathbf{e}_k$$

since the starting capital is k, with certainty.

The following result, which relates the transition matrix P to the probability density vectors, is fundamental.

Theorem 11.1. *If P denotes the transition matrix of the game (see (11.1)), then the probability density vectors satisfy the equation*

$$\pi^{(t+1)} = P\pi^{(t)}, \quad t = 0, 1, \ldots \tag{11.2}$$

Proof: Let us first assume $2 \leq j \leq N-2$ and let us try to compute the j–th component of $\pi^{(t+1)}$. There are two independent possibilities to have capital j at time $t+1$: We can have capital $j-1$ at time t and win (with probability p) or we can have capital $j+1$ at time t and lose (with probability q). This yields:

$$\begin{aligned} \pi_j^{(t+1)} &= prob\Big(X_{t+1} = j\Big) \\ &= p \cdot prob\Big(X_t = j-1\Big) + q \cdot prob\Big(X_t = j+1\Big) \\ &= p \cdot \pi_{j-1}^{(t)} + q \cdot \pi_{j+1}^{(t)} \end{aligned}$$

Now note that the j–th row of P is

$$(0 \ldots 0 \; p \; 0 \; q \; 0 \ldots 0)$$

where the centered 0 is in position j. Therefore,

$$(P\pi^{(t)})_j = p \cdot \pi_{j-1}^{(t)} + q \cdot \pi_{j+1}^{(t)}$$

and we have shown that

$$\pi_j^{(t+1)} = (P\pi^{(t)})_j$$

for $2 \leq j \leq N-2$.

Next let $j = 1$. How can we reach the capital $j = 1$ at time $t+1$? There is only one possibility, namely to have capital 2 at time t and losing (with probability q). (Note that $X_t = 0$ leads to $X_{t+1} = 0$ by Rule 1 of Section 1..) This yields:

$$\begin{aligned} \pi_1^{(t+1)} &= prob\Big(X_{t+1} = 1\Big) \\ &= q \cdot prob\Big(X_t = 2\Big) \\ &= q \cdot \pi_2^{(t)} \\ &= (P\pi^{(t)})_1 \end{aligned}$$

The last equation holds since row 1 of the transition matrix P is

$$(0 \; 0 \; q \; 0 \; \ldots \; 0)$$

with q in position 2. For $j = 0, j = N-1, j = N$ the equation

$$\pi_j^{(t+1)} = (P\pi^{(t)})_j$$

follows similarly. This proves the vector equation (11.2). $\diamond$

5. Discussion

Why is equation (11.2) interesting? Most remarkable is this: We started with a *random* evolution $X_t \to X_{t+1}$, specified by the Rules 1 to 3 in Section 1., and obtained a *deterministic* evolution (11.2)! Of course, what evolves deterministically is *not* the random variable X_t, but its probability density vector $\pi^{(t)}$. For random evolution, this is generally the best one can hope for: *To get a deterministic law which shows how probabilities evolve.*

In quantum physics, one uses the Schrödinger equation for the wave function $\psi(x,t)$ to describe how the quantum state of a physical system changes in time. The Schrödinger equation is deterministic and the wave function $\psi(x,t)$ evolves deterministically. Since $\psi(x,t)$ is complex valued, it cannot directly equal probability density. In fact, the function $|\psi(\cdot,t)|^2$ can be identified with probability density and it evolves deterministically.

6. Application

Let us give a concrete application of formula (11.2). As a simple example, consider the case where the maximal capital is $N = 4$, the starting capital is $k = 2$, the probability of winning (in each step) is $p = 0.4$ and the probability of losing is $q = 1 - p = 0.6$. Because $p < q$, our intuition tells us that the gambler is more likely to get ruined, ending with zero capital, $X_t = 0$, than to win the game and end with $X_t = 4$. This intuition is correct, but using formula (11.2) we are now able to quantify our intuition.

The initial probability density vector is

$$\pi^{(0)} = \begin{pmatrix} 0 \\ 0 \\ 1 \\ 0 \\ 0 \end{pmatrix}$$

and the transition matrix is

$$P = \begin{pmatrix} 1 & 0.6 & 0 & 0 & 0 \\ 0 & 0 & 0.6 & 0 & 0 \\ 0 & 0.4 & 0 & 0.6 & 0 \\ 0 & 0 & 0.4 & 0 & 0 \\ 0 & 0 & 0 & 0.4 & 1 \end{pmatrix}.$$

Using Matlab (see the script given below), it is easy to compute the vectors

$$\pi^{(n)} = P^n \pi^{(0)}, \quad n = 0, 1, 2, \ldots$$

(We write n here instead of t to emphasize that time is an integer.)

Our output tells us, for example, that (up to round–off)

$$\pi^{(10)} = \begin{pmatrix} 0.6747 \\ 0 \\ 0.0255 \\ 0 \\ 0.2999 \end{pmatrix}$$

What does this mean? After 10 games, the probability to have capital $X_{10} = 0$ is

$$\pi_0^{(10)} = 0.6747$$

and the probability to have capital $X_{10} = 4$ is

$$\pi_4^{(10)} = 0.2999$$

There is still the small probability

$$\pi_2^{(10)} = 0.0255$$

that the game is not over and, after 10 games, the gambler still has the capital $X_{10} = 2$. (It is easy to see that, after an even number of games and using an even starting capital, the probability to have an odd capital is exactly zero.)

Computing $\pi^{(n)} = P^n \pi^{(0)}$ for increasing n, one obtains numerically that $\pi^{(n)}$ approaches the vector

$$\pi^{(\infty)} = \begin{pmatrix} 0.6923 \\ 0 \\ 0 \\ 0 \\ 0.3077 \end{pmatrix}.$$

The last component of this vector is the probability to win the game and to go home with $X_t = 4$ dollars.

7. Script: Evolving the Probability Density for Gambler's Ruin

```
%%%%%%%%%%%%%%%%%%%%%%%%%%%%%%%%%%%%%%%%%%%%%%%%%%%%%%%%%%%%%%
% g4.m
% Evolution of probability density vector for gambler's
% ruin
%%%%%%%%%%%%%%%%%%%%%%%%%%%%%%%%%%%%%%%%%%%%%%%%%%%%%%%%%%%%%%
clear
```

```
% specification of parameters
p=0.4;
q=1-p;
N=4;
k=2;
nmax=10;
%%%%%%%%%%%%%%%%%%%%%%%%%%%%%%%%%%%%%%%%%%%%%%%%%%%%%%%%%%
% The matrix P
%%%%%%%%%%%%%%%%%%%%%%%%%%%%%%%%%%%%%%%%%%%%%%%%%%%%%%%%%%
NN=N+1;
P=zeros(NN);
P(1,1)=1;
P(NN,NN)=1;
for j=2:N
P(j-1,j)=q;
P(j+1,j)=p;
end
%%%%%%%%%%%%%%%%%%%%%%%%%%%%%%%%%%%%%%%%%%%%%%%%%%%%%%%%%%
% The starting vector x
%%%%%%%%%%%%%%%%%%%%%%%%%%%%%%%%%%%%%%%%%%%%%%%%%%%%%%%%%%
for j=1:NN
x(j)=0;
end
x(k+1)=1;
x=x';
%%%%%%%%%%%%%%%%%%%%%%%%%%%%%%%%%%%%%%%%%%%%%%%%%%%%%%%%%%
% evolution of x for nmax steps
% the vector P^nx is stored as column n+1 of the matrix Z
%%%%%%%%%%%%%%%%%%%%%%%%%%%%%%%%%%%%%%%%%%%%%%%%%%%%%%%%%%
Z(:,1)=x;
for n=1:nmax
x=P*x;
Z(:,n+1)=x;
end
```

Chapter 12

Gambler's Ruin: Probabilities and Expected Time

Summary: Recall from the previous chapter that $X_t \in \{0, 1, \ldots, N\}$ denotes the capital of the gambler at time t, and time progresses in integer steps, $t = 0, 1, 2, \ldots$

The initial capital is

$$X_0 = k \in \{0, 1, \ldots, N\}$$

and X_t evolves randomly: If $1 \leq X_t \leq N - 1$ then $X_{t+1} = X_t + 1$ with probability p (winning), $X_{t+1} = X_t - 1$ with probability $q = 1 - p$ (losing). If $X_t = 0$ or $X_t = N$ then $X_{t+1} = X_t$. The game has ended.

In this chapter, we will determine the probability α_k for ruin to occur if the initial capital is $X_0 = k$. Ruin means that we have $X_t = 0$ at some time t in the future. Interestingly, we do not determine the probabilities α_k for one index k at a time, but will determine the α–vector in one process. Similarly, we determine the probability β_k of ultimately winning, i.e., of $X_t = N$ to occur at some time t in the future. We will also determine the expected length τ_k of the game.

In the last section of this chapter, we reconsider the evolution of the probability density vectors $P^n \mathbf{e}_k$, which were introduced in Section 4.. If the initial capital is $X_0 = k$, then the vector $P^n \mathbf{e}_k$ is the probability density vector at time $t = n$, i.e., the component $(P^n \mathbf{e}_k)_j$ gives us the probability to have capital j at time $t = n$. We have computed the vectors $P^n \mathbf{e}_k$ numerically in Section 6. and have observed, in an example, that they approach a limit as time $t = n$ tends to infinity. We will prove here that the limit $\lim_{n\to\infty} P^n \mathbf{e}_k$ always exists and is determined by α_k and β_k. Some basic knowledge of linear algebra is required to understand the arguments.

1. Probability of Ruin

The maximal capital N as well as the probabilities p and $q = 1 - p$ are fixed. $X_0 = k$ denotes the starting capital. If α_k denotes the probability for $X_t = 0$ to occur at some time t, then our rules imply

$$\alpha_0 = 1 \quad \text{and} \quad \alpha_N = 0 \; . \tag{12.1}$$

This simply says that if our initial capital is $X_0 = 0$ we will surely be ruined and if $X_0 = N$ we will surely not get ruined.

Now consider the less trivial case where $1 \le X_0 = k \le N - 1$. After one game we have

$$X_1 = \begin{cases} k+1 & \text{with probability } p \\ k-1 & \text{with probability } q \end{cases} .$$

If the capital is $k \pm 1$ then the probability of ruin to occur at some time in the future equals $\alpha_{k\pm1}$ and we conclude that

$$\alpha_k = p\alpha_{k+1} + q\alpha_{k-1} \quad \text{for} \quad 1 \le k \le N-1 \; .$$

We will now solve the homogeneous difference equation

$$p\alpha_{k+1} - \alpha_k + q\alpha_{k-1} = 0 \quad \text{for} \quad 1 \le k \le N-1 \tag{12.2}$$

with the boundary conditions (12.1).

To solve the difference equation, use the ansatz

$$\alpha_k = \rho^k, \quad k = 0, 1, \ldots$$

The unknown ρ is raised to its k–th power. For ρ one obtains the quadratic

$$p\rho^2 - \rho + q = 0 \; . \tag{12.3}$$

Case 1, $p \neq \frac{1}{2}$: In this case, the quadratic has the distinct roots

$$\rho_1 = 1, \quad \rho_2 = q/p$$

as one easily checks. Therefore, the linear, homogeneous difference equations (12.2) have the general solution

$$\alpha_k = c_1 + c_2(q/p)^k, \quad k = 0, 1, 2, \ldots \tag{12.4}$$

with free constants c_1, c_2. These constants are determined by the boundary conditions (12.1). One obtains:

$$\begin{aligned} 1 = \alpha_0 &= c_1 + c_2 \ , \\ 0 = \alpha_N &= c_1 + c_2 (q/p)^N \ . \end{aligned}$$

This linear system for $c_{1,2}$ is easy to solve and equation (12.4) yields

$$\alpha_k = \frac{1-(q/p)^{k-N}}{1-(q/p)^{-N}} \quad \text{for} \quad 0 \le k \le N \ .$$

Example: Let $N = 4, p = 0.4, q = 0.6, k = 2$. One finds that

$$\alpha_2 = \frac{1-(3/2)^{-2}}{1-(3/2)^{-4}} = 0.6923$$

This value agrees with the 0–component of the vector $\pi^{(\infty)}$, which was computed numerically in Section 6..

Case 2, $p = q = \frac{1}{2}$: In this case, the quadratic (12.3) has the double root

$$\rho_1 = \rho_2 = 1$$

and the general solution of the difference equations (12.2) is

$$\alpha_k = c_1 + c_2 k \ .$$

Imposing the boundary conditions $\alpha_0 = 1, \alpha_N = 0$ yields the simple formula

$$\alpha_k = 1 - \frac{k}{N}, \quad 0 \le k \le N \ .$$

2. Probability of Winning

Let β_k denote the probability to ultimately win, i.e., to reach $X_t = N$ at some time t, assuming the initial condition $X_0 = k$. We proceed in the same way as for the α_k. Only the boundary condition (12.1) must be changed. They become

$$\beta_0 = 0 \quad \text{and} \quad \beta_N = 1 \tag{12.5}$$

since winning cannot occur if $X_0 = 0$ and winning is certain if $X_0 = N$. Proceeding as in Section 1. one obtains that

$$\beta_k = \frac{(q/p)^{k-N} - (q/p)^{-N}}{1-(q/p)^{-N}} \quad \text{for} \quad 0 \le k \le N$$

if $p \neq \frac{1}{2}$ and

$$\beta_k = \frac{k}{N} \quad \text{for} \quad 0 \le k \le N$$

if $p = q = \frac{1}{2}$.

Note that $\alpha_k + \beta_k = 1$ for every k. Thus, the probability to reach either $X_t = 0$ or $X_t = N$ equals one. With probability one, either ruin or ultimate winning will occur. We note, however, that it is *possible* for the game to go on forever. Just assume the coin tosses show heads, then tails, then heads, then tails, etc., going on forever in this alternating way. This is *possible*, but has zero probability.

3. Expected Time

As above, we fix the maximal capital N and the probabilities p and $q = 1 - p$. If we run the game starting with $X_0 = k$ then, with probability one, we will reach $X_t = 0$ or $X_t = N$ at some time t in the future. If $t = t_0$ is the *smallest* time with $X_{t_0} = 0$ or $X_{t_0} = N$, then we call t_0 the length of the game. Of courses, this length t_0 will not only depend on k, but on the whole realization X_t of the random evolution, i.e., on the outcome of the coin tosses. We now fix k, run many realizations, and take the average of all observed lengths, t_0. Intuitively, this average is the *expected time* for the game with starting value $X_0 = k$. We denote the expected time by τ_k.

Clearly,

$$\tau_0 = \tau_N = 0 \; . \tag{12.6}$$

If $X_0 = 0$ or $X_0 = N$, then *every* realization of the game has length zero and, of course, the expected time for the game is zero.

Recall that $1 \le X_0 = k \le N - 1$ yields

$$X_1 = \begin{cases} k+1 & \text{with probability } p \; , \\ k-1 & \text{with probability } q \; . \end{cases}$$

For example, with probability p we have $X_1 = k + 1$ and then expect the game to last for *the additional time* τ_{k+1}. Therefore,

$$\tau_k = p(1 + \tau_{k+1}) + q(1 + \tau_{k-1}) \quad \text{for} \quad 1 \le k \le N - 1 \; .$$

The number 1 is added to τ_{k+1} and to τ_{k-1} in the above formula, because one game has been played to arrive at $X_1 = k + 1$ or $X_1 = k - 1$. Since $p + q = 1$ the above formula can be rewritten as the inhomogeneous difference equations

$$p\tau_{k+1} - \tau_k + q\tau_{k-1} = -1 \quad \text{for} \quad 1 \le k \le N - 1 \; , \tag{12.7}$$

which we must solve subject to the boundary conditions (12.6).

The solution process is similar to solving linear inhomogeneous differential equations, which may be more familiar to the reader. One guesses a particular solution of the inhomogeneous equation and adds the general solution of the homogeneous equation to obtain the general solution of the inhomogeneous equation.

First, we set

$$\tilde{\tau}_k = k, \quad k = 0, 1, \ldots, N .$$

and calculate

$$\begin{aligned} p\tilde{\tau}_{k+1} - \tilde{\tau}_k + q\tilde{\tau}_{k-1} &= p(k+1) - k + q(k-1) \\ &= p - q \end{aligned}$$

It is important to note that the result, $p - q$, does not depend on k. Therefore,

$$\tau_k^{(part)} = \frac{k}{q-p}, \quad 0, 1, \ldots, N ,$$

is a particular solution of the inhomogeneous difference equations (12.7). Now let's assume, for definiteness, that $p \neq \frac{1}{2}$. We add the general solution of the homogeneous difference equation to $\tau_k^{(part)}$:

$$\tau_k = \frac{k}{q-p} + c_1 + c_2(q/p)^k, \quad 0 \le k \le N ,$$

with free constants $c_{1,2}$. This is the general solution of the inhomogeneous equations (12.7). The boundary conditions (12.6) fix the constants $c_{1,2}$.

One then obtains for the expected times

$$\tau_k = \frac{k}{q-p} - \frac{N}{q-p} \cdot \frac{(q/p)^{k-N} - (q/p)^{-N}}{1 - (q/p)^{-N}}, \quad k = 0, 1, \ldots, N . \tag{12.8}$$

4. The Matrix View: Limit of Probability Densities

In Sections 1. and 2. we have determined the vectors α and β whose components α_k and β_k are the probabilities of ruin and ultimate winning if the initial capital is $X_0 = k$. In this section, we want to look at these vectors using the transition matrix P, which was introduced in Section 3..

The matrix P is important if one wants to study how probability density vectors evolve in time. If $X_0 = k$ is the initial capital, then the probability density vector at time $t = n$ is $P^n \mathbf{e}_k$. Our main result in this section says that the probability density vectors $P^n \mathbf{e}_k$ approach a limit as $n \to \infty$, and the limit depends on α_k and β_k. Precisely:

Theorem 12.1. *For every $0 \le k \le N$, the sequence of probability density vectors $P^n \mathbf{e}_k$ converges and the limit is*

$$\lim_{n\to\infty} P^n \mathbf{e}_k = \begin{pmatrix} \alpha_k \\ 0 \\ \vdots \\ 0 \\ \beta_k \end{pmatrix} . \qquad (12.9)$$

4.1. The Eigenvalues of P

If $N = 4$ we have

$$P = \begin{pmatrix} 1 & q & 0 & \dots & 0 \\ 0 & 0 & q & & 0 \\ 0 & p & 0 & q & 0 \\ \vdots & & \ddots & \ddots & 0 \\ 0 & \dots & 0 & p & 1 \end{pmatrix}, \quad P^T = \begin{pmatrix} 1 & 0 & 0 & \dots & 0 \\ q & 0 & p & & 0 \\ 0 & \ddots & \ddots & \ddots & 0 \\ \vdots & & q & 0 & p \\ 0 & \dots & 0 & 0 & 1 \end{pmatrix} .$$

One easily checks that the vectors

$$\alpha = \begin{pmatrix} 1 \\ \alpha_1 \\ \vdots \\ \alpha_{N-1} \\ 0 \end{pmatrix} \quad \text{and} \quad \beta = \begin{pmatrix} 0 \\ \beta_1 \\ \vdots \\ \beta_{N-1} \\ 1 \end{pmatrix}$$

satisfy

$$P^T \alpha = \alpha \quad \text{and} \quad P^T \beta = \beta . \qquad (12.10)$$

In terms of linear algebra, this says that α and β are eigenvectors of P^T for the eigenvalue $\lambda = 1$. Therefore, the matrix P also has $\lambda = 1$ as a double eigenvalue. It is easy to check this directly since

$$P\mathbf{e}_0 = \mathbf{e}_0, \quad P\mathbf{e}_N = \mathbf{e}_N .$$

Let us determine the eigenvalues of P^T which are different from 1 and assume that

$$P^T \phi = \lambda \phi, \quad \phi \in \mathbb{C}^{N+1}, \quad \phi \neq 0, \quad \lambda \neq 1 .$$

Since

$$(P^T\phi)_0 = \phi_0 \quad \text{and} \quad (P^T\phi)_N = \phi_N$$

and $\lambda \neq 1$ it follows that

$$\phi_0 = \phi_N = 0 \ .$$

Let Q denotes the $(N-1) \times (N-1)$ matrix which one obtains by deleting the rows and columns of P^T with indices zero and N:

$$Q = \begin{pmatrix} 0 & p & & & 0 \\ q & 0 & p & & \\ & \ddots & \ddots & \ddots & \\ & & q & 0 & p \\ 0 & & & q & 0 \end{pmatrix} .$$

Also, let

$$\tilde{\phi} = \begin{pmatrix} \phi_1 \\ \vdots \\ \phi_{N-1} \end{pmatrix} . \tag{12.11}$$

It is then clear that

$$Q\tilde{\phi} = \lambda\tilde{\phi} \ .$$

In other words, the eigenvalues of P^T which are different from one are the eigenvalues of Q.

We now symmetrize Q by a similarity transformation with a diagonal matrix D. If

$$D := diag(d, d^2, \ldots, d^{N-1}), \quad d > 0 \ ,$$

then

$$D^{-1}QD = \begin{pmatrix} 0 & pd & & & 0 \\ q/d & 0 & pd & & \\ & \ddots & \ddots & \ddots & \\ & & q/d & 0 & pd \\ 0 & & & q/d & 0 \end{pmatrix} .$$

We choose $d > 0$ so that $pd = q/d$, i.e., $d = \sqrt{q/p}$. For this choice of d we have

$$pd = q/d = \sqrt{pq}$$

and

$$D^{-1}QD = \sqrt{pq}\, S$$

where S is the $(N-1) \times (N-1)$ matrix

$$S = \begin{pmatrix} 0 & 1 & & & 0 \\ 1 & 0 & 1 & & \\ & \ddots & \ddots & \ddots & \\ & & 1 & 0 & 1 \\ 0 & & & 1 & 0 \end{pmatrix} .$$

Lemma 12.1. *The $(N-1) \times (N-1)$ matrix S has the eigenvalues*

$$\mu_k = 2\cos(\pi k), \quad 1 \le k \le N-1 .$$

Proof: Fix $1 \le k \le N-1$ and set

$$\phi_j = \sin(\pi j k), \quad 0 \le j \le N .$$

Then $\phi_0 = \phi_N = 0$ and, for $1 \le j \le N-1$:

$$\begin{aligned} \phi_{j-1} + \phi_{j+1} &= \sin(\pi(j-1)k) + \sin(\pi(j+1)k) \\ &= 2\cos(\pi k)\sin(\pi j k) \\ &= 2\mu_k \phi_j \end{aligned}$$

This yields that the corresponding vector $\tilde{\phi}$ (see (12.11)) is an eigenvector of S to the eigenvalue μ_k. ◇

Since Q is similar to $\sqrt{pq}\, S$ we obtain the following result:

Theorem 12.2. *The $(N+1) \times (N+1)$ matrix P^T has the eigenvalues $\lambda_0 = \lambda_N = 1$ and*

$$\lambda_k = 2\sqrt{pq}\, \cos(\pi k), \quad 1 \le k \le N-1 . \tag{12.12}$$

The function $p \to pq = p(1-p)$ has the maximal value $\frac{1}{4}$. Therefore, the eigenvalues λ_k given in (12.12) are all strictly less than one in absolute value.

Corollary: *The $(N+1) \times (N+1)$ matrix P has the double eigenvalue one,*

$$P\mathbf{e}_0 = \mathbf{e}_0, \quad P\mathbf{e}_N = \mathbf{e}_N .$$

The other eigenvalues of P are the $N-1$ numbers λ_k given in (12.12), which are strictly less than one in absolute value.

This Corollary will be important to obtain convergence of $P^n y$ as $n \to \infty$, as we will see below.

4.2. Convergence of Probability Density Vectors

First let $y \in \mathbb{R}^{N+1}$ denote an arbitrary vector and consider the sequence

$$P^n y, \quad n = 0, 1, 2, \dots$$

We will show that this sequence always converges as $n \to \infty$. We will then take $y = \mathbf{e}_k$. In this case, the vectors $P^n \mathbf{e}_k$ are the probability density vectors which we considered in Section 4..

According to the above Corollary, the space $\mathbb{R}^{N+1}$ has a basis consisting of eigenvectors of P, which we may choose as

$$\mathbf{e}_0, \quad \mathbf{e}_N, \quad \mathbf{s}_1, \dots, \mathbf{s}_{N-1}$$

with

$$P\mathbf{e}_0 = \mathbf{e}_0, \quad P\mathbf{e}_N = \mathbf{e}_N, \quad P\mathbf{s}_k = \lambda_k \mathbf{s}_k, \quad |\lambda_k| < 1 \; .$$

We write the given vector $y \in \mathbb{R}^{N+1}$ as

$$y = c_0 \mathbf{e}_0 + c_N \mathbf{e}_N + \sum_{k=1}^{N-1} c_k \mathbf{s}_k \; .$$

Then application of P^n yields

$$P^n y = c_0 \mathbf{e}_0 + c_N \mathbf{e}_N + \sum_{k=1}^{N-1} c_k \lambda_k^n \mathbf{s}_k$$

and, since $\lambda_k^n \to 0$ as $n \to 0$, we obtain

$$P^n y \to c_0 \mathbf{e}_0 + c_N \mathbf{e}_N \quad \text{as} \quad n \to \infty \; .$$

To determine the limit of $P^n \mathbf{e}_k$, we use the inner product

$$\langle a, b \rangle = \sum_{j=0}^{N} a_j b_j$$

on $\mathbb{R}^{N+1}$ and recall the important rule

$$\langle a, Pb \rangle = \langle P^T a, b \rangle \; .$$

Since $P^T \alpha = \alpha$ we have

$$\begin{aligned}\langle \alpha, P^n \mathbf{e}_k \rangle &= \langle (P^T)^n \alpha, \mathbf{e}_k \rangle \\ &= \langle \alpha, \mathbf{e}_k \rangle \\ &= \alpha_k\end{aligned}$$

for every n. Taking the inner product with α in the limit relation

$$P^n \mathbf{e}_k \to c_0 \mathbf{e}_0 + c_N \mathbf{e}_N \quad \text{as} \quad n \to \infty \tag{12.13}$$

we obtain

$$\alpha_k = \langle \alpha, P^n \mathbf{e}_k \rangle \to c_0 \langle \alpha, \mathbf{e}_0 \rangle + c_N \langle \alpha, \mathbf{e}_N \rangle \ .$$

We have seen above that the vector α has the form

$$\alpha = (1, \alpha_1, \ldots, \alpha_{N-1}, 0)^T \ ,$$

which implies that

$$\langle \alpha, \mathbf{e}_0 \rangle = 1 \quad \text{and} \quad \langle \alpha, \mathbf{e}_N \rangle = 0 \ .$$

Therefore, the coefficient c_0 in the limit relation (12.13) equals

$$c_0 = \alpha_k \ .$$

In the same way, one shows that

$$c_N = \beta_k \ .$$

We have proved our main result on the limit of probability density vectors, Theorem 12.1.

Chapter 13

Stochastic Model of a Simple Growth Process

Summary: In this chapter, we consider the simplest model for a stochastic growth process in continuous time t. The random variable X_t, a positive integer, denotes the number of individuals in the population at time t. At some random times

$$T_1 < T_2 < T_3 \ldots$$

a birth occurs, and X_t increases by one. The expected value and the variance of X_t as well as the statistics of interevent times will be calculated.

1. Growth Models

First, let us briefly recall the derivation of the simplest *deterministic* growth model, leading to the ODE $u'(t) = bu(t)$. If $u(t)$ denotes the size of a population at time t and $b > 0$ denotes the birth rate per individual then, for small $\Delta t > 0$,

$$u(t + \Delta t) \sim u(t) + u(t) b \Delta t \; .$$

The term $u(t)b\Delta t$ accounts for population growth in the time interval from t to $t + \Delta t$. One obtains

$$\frac{1}{\Delta t}\Big(u(t + \Delta t) - u(t)\Big) \sim bu(t)$$

and, for $\Delta t \to 0$,

$$u'(t) = bu(t) \; .$$

An initial population $u(0) = k$ grows exponentially, $u(t) = ke^{bt}$. This growth model can be criticized for many reasons. As we noted earlier, exponential growth cannot go on for very long, and logistic growth may be somewhat more realistic. In this chapter, however, we will consider a completely different modification and replace the population size $u(t)$ by a random variable X_t. Here $X_t \in \{1, 2, 3, \ldots\}$ denotes the number of individuals in the population at time t. For large populations, it is appropriate to model the population size by a *real* variable, but for small population sizes — a few dozen, say — it is often better to use an integer which counts the number of individuals in the population.

How shall we model the evolution of the random variable X_t in the simplest case? In Section 2. below we will consider the probabilities

$$p_j(t) = prob(X_t = j), \quad j = 1, 2, \ldots$$

i.e., $p_j(t)$ is the probability that the population size equals j at time t. In our model, we will assume that $X_0 = k$ is a given positive integer, the number of individuals at time $t = 0$. The probabilities $p_j(t)$ will depend on the initial value $X_0 = k$, of course, but we will suppress this dependency in our notation. In Section 2. we will derive the so–called *Forward Kolmogorov Equation* for the functions $p_j(t)$, a system of ODEs. Together with the initial condition

$$p_j(0) = \delta_{jk}, \quad j = 1, 2, \ldots$$

these equations determine how the probabilities $p_j(t)$ evolve *deterministically*. In Section 3. we will solve the Forward Kolmogorov Equations and obtain explicit expressions for the $p_j(t)$. As an exercise, and a check that our result is not absurd, we will compute the series $\sum_j p_j(t)$ and show that — at each time t — it equals one; see Section 4.. This is, of course, how it should be because the random variable X_t takes on exactly one of the values $j \in \{1, 2, \ldots\}$.

We will then use the probabilities $p_j(t)$ to determine the expected value and the variance of the random variable X_t. Finally, we will calculate the probability distribution of interevent times, i.e., of the times between consecutive births. This allows us to run numerical realizations for the evolution of the random variable X_t.

2. The Forward Kolmogorov Equations

In this model, time is continuous, $0 \le t < \infty$, and the state space is discrete,

$$A = \mathbb{N} = \{1, 2, 3, \ldots\} \ .$$

With

$$X_t \in \mathbb{N}$$

we denote the number of individuals in the population at time t. Clearly, X_t must be an integer. In large populations the variable X_t is often treated as a continuous variable, but in small populations it is better to treat X_t as an integer.

We let

$$p_j(t) := prob\Big(X_t = j\Big), \quad j \in \mathbb{N} \ .$$

Let us assume that at time $t = 0$ the population has k individuals,

$$X_0 = k \in \mathbb{N} \ ,$$

thus

$$p_j(0) = prob\Big(X_0 = j\Big) = \delta_{jk}, \quad j \in \mathbb{N} \ .$$

In our model, no deaths occur and the probability for an individual to give birth in a time interval of length Δt is

$$b\Delta t$$

for small Δt. We choose Δt so small that *no* birth or *precisely one* birth will occur in the time interval from t to $t + \Delta t$. We want to obtain an equation for

$$p_j(t + \Delta t) = prob\Big(X_{t+\Delta t} = j\Big) \ .$$

There are two independent possibilities for $X_{t+\Delta t}$ to equal j: First, it is possible that at time t there were $j - 1$ individuals and a birth occurred in the interval from t to $t + \Delta t$. Second, at time t there were j individuals and no birth occurred in the interval from t to $t + \Delta t$.

Recall that $b > 0$ denotes the birth rate per individual. Then the probability for the first case is

$$p_{j-1}(t)b(j-1)\Delta t \ .$$

The probability of the second case is

$$p_j(t)(1 - bj\Delta t) \ .$$

To arrive at the formula for the second case, note that the probability for a birth to occur in the interval from t to $t + \Delta t$ is $bj\Delta t$ if one assumes that at time t there are j individuals in the population.

These assumptions yield

$$p_j(t + \Delta t) = p_{j-1}(t)b(j-1)\Delta t + p_j(t)(1 - bj\Delta t) \ .$$

We can rewrite this equation as

$$\frac{1}{\Delta t}\Big(p_j(t+\Delta t) - p_j(t)\Big) = p_{j-1}(t)b(j-1) - bjp_j(t) \ .$$

For $\Delta t \to 0$ we obtain

$$p_j'(t) = -bjp_j(t) + b(j-1)p_{j-1}(t), \quad j = 1, 2, \ldots \tag{13.1}$$

with $p_0(t) \equiv 0$. We have derived the so–called *Forward Kolmogorov Equations* of the model. They from an infinite system of ODEs:

$$\begin{aligned}
p_1' &= -bp_1 \\
p_2' &= -2bp_2 + bp_1 \\
p_3' &= -3bp_3 + 2bp_2 \\
p_4' &= -4bp_4 + 3bp_3
\end{aligned}$$

etc.

3. Solution of the Forward Kolmogorov Equations

In this section we will solve the equations (13.1) with the initial condition

$$p_j(0) = \delta_{jk}, \quad j = 1, 2, \ldots$$

where $k = X_0 \geq 1$ is the size of the population at time $t = 0$. We obtain

$$p_j(t) \equiv 0, \quad j < k \ ,$$

and

$$p_k' = -kbp_k, \quad p_k(0) = 1 \ ,$$

thus

$$p_k(t) = e^{-kbt} \ .$$

It is clear that, in principle, we can successively determine $p_{k+1}(t), p_{k+2}(t), \ldots$ by solving ODE initial value problems.

Interestingly, it is not difficult to derive closed form expressions for the $p_j(t)$. First, with new unknown functions $\phi_j(t)$ let

$$p_j(t) = e^{-jbt}\phi_j(t) \ . \tag{13.2}$$

As we will now show, this transformation eliminates the diagonal term in the ODE system (13.1) for the p_j. Using (13.2) one obtains

$$p_j' = -jbp_j + e^{-jbt}\phi_j' \ ,$$

and (13.1) becomes

$$e^{-jbt}\phi_j' = b(j-1)e^{-(j-1)bt}\phi_{j-1} \ ,$$

thus

$$\phi_j' = (j-1)be^{bt}\phi_{j-1} \ .$$

If we compare the above system for the functions $\phi_j(t)$ with the system (13.1) for the functions $p_j(t)$, we see that the diagonal terms are eliminated.

Together with the initial conditions $\phi_j(0) = \delta_{jk}$ (for all j) we find that

$$\phi_k \equiv 1$$

and

$$\phi_j' = (j-1)be^{bt}\phi_{j-1}, \quad \phi_j(0) = 0 \quad \text{for} \quad j \geq k+1 \ .$$

In principle, the ϕ_j can now be obtained by successive integrations.

The following turns out to work: With unknown coefficients a_j let

$$\begin{aligned} \phi_j(t) &= a_j(e^{bt}-1)^{j-k} \\ \phi_{j-1}(t) &= a_{j-1}(e^{bt}-1)^{j-1-k} \ . \end{aligned}$$

The formula for $\phi_j(t)$ yields

$$\phi_j'(t) = a_j(j-k)(e^{bt}-1)^{j-1-k}be^{bt}$$

and the differential equation $\phi_j' = (j-1)be^{bt}\phi_{j-1}$ together with the formula for ϕ_{j-1} yields

$$\phi_j'(t) = (j-1)be^{bt}a_{j-1}(e^{bt}-1)^{j-1-k} \ .$$

One thus obtains the following conditions for the unknown coefficients a_j:

$$a_j(j-k) = a_{j-1}(j-1) \quad \text{for} \quad j \geq k+1, \quad a_k = 1 \ . \tag{13.3}$$

The equations

$$a_j = \frac{j-1}{j-k}a_{j-1}, \quad j \geq k+1, \quad a_k = 1$$

imply

$$\begin{aligned} a_{k+1} &= \frac{k}{1} \\ a_{k+2} &= \frac{k+1}{2} \cdot \frac{k}{1} \\ a_{k+3} &= \frac{k+2}{3} \cdot \frac{k+1}{2} \cdot \frac{k}{1} \end{aligned}$$

etc. The a_j can be written as binomial coefficients:

$$a_j = \frac{(j-1)!}{(k-1)!(j-k)!} = \begin{pmatrix} j-1 \\ k-1 \end{pmatrix} \quad \text{for} \quad j \geq k \ ,$$

and one obtains the closed form solution

$$p_j(t) = \begin{pmatrix} j-1 \\ k-1 \end{pmatrix} e^{-jbt}(e^{bt}-1)^{j-k} \quad \text{for} \quad j \geq k \ ,$$

of the Forward Kolmogorov Equations with initial condition $p_j(0) = \delta_{jk}$.

In Figure 13.1 we show the functions $p_1(t), p_2(t)$, and $p_3(t)$ for $b = k = 1$. Note that the choice $b = 1$ is not restrictive since it can always be enforced by choosing a suitable unit of time. The choice $k = 1$ may correspond to a single cell which can grow into many cells by successive cell divisions.

The graph of $p_2(t)$ shows, for example, that the probability to have precisely two cells, corresponding to $X_t = 2$, is maximal at time $t \sim 0.7$.

It is also interesting to plot the sequence

$$p_j(t), \quad j = 1, 2, 3, \ldots$$

for fixed t.

In Figure 13.2 we show the probabilities $p_j(1)$ and $p_j(2)$ as functions of j where $b = 1, k = 2$ are fixed. We see, for example, that at time $t = 1$ the value of X_1 with highest probability is $j = 3$; its probability is about $p_3(1) \sim 0.17$. At time $t = 2$ the value of X_2 with highest probability is $j = 7$.

4. The Sum of the $p_j(t)$

In this section, we want to check that

$$\sum_{j=k}^{\infty} p_j(t) = 1 \tag{13.4}$$

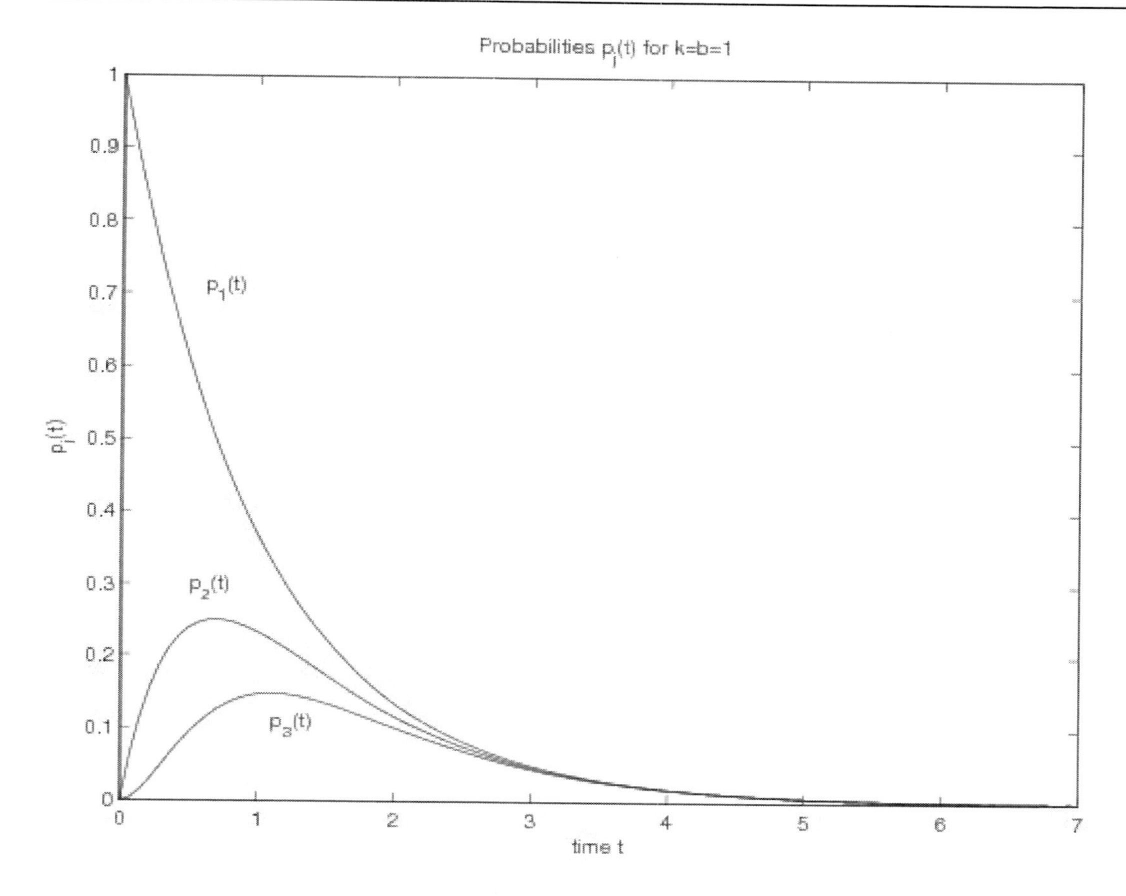

Figure 13.1. Time evolution of $p_j(t)$ for $b = k = 1$

where the $p_j(t)$ are the solutions of the Forward Kolmogorov Equations determined in the previous section. The technique to show (13.4) will be useful also in the following sections, where we will compute the expected value and the variance of the random variable X_t.

Let us first give a formal argument for (13.4), which uses only the Forward Kolmogorov Equations (13.1) and the initial condition $p_j(0) = \delta_{jk}$. If we define the function

$$p(t) = \sum_{j=k}^{\infty} p_j(t)$$

then $p(0) = 1$ and, formally differentiating the series term by term, we obtain

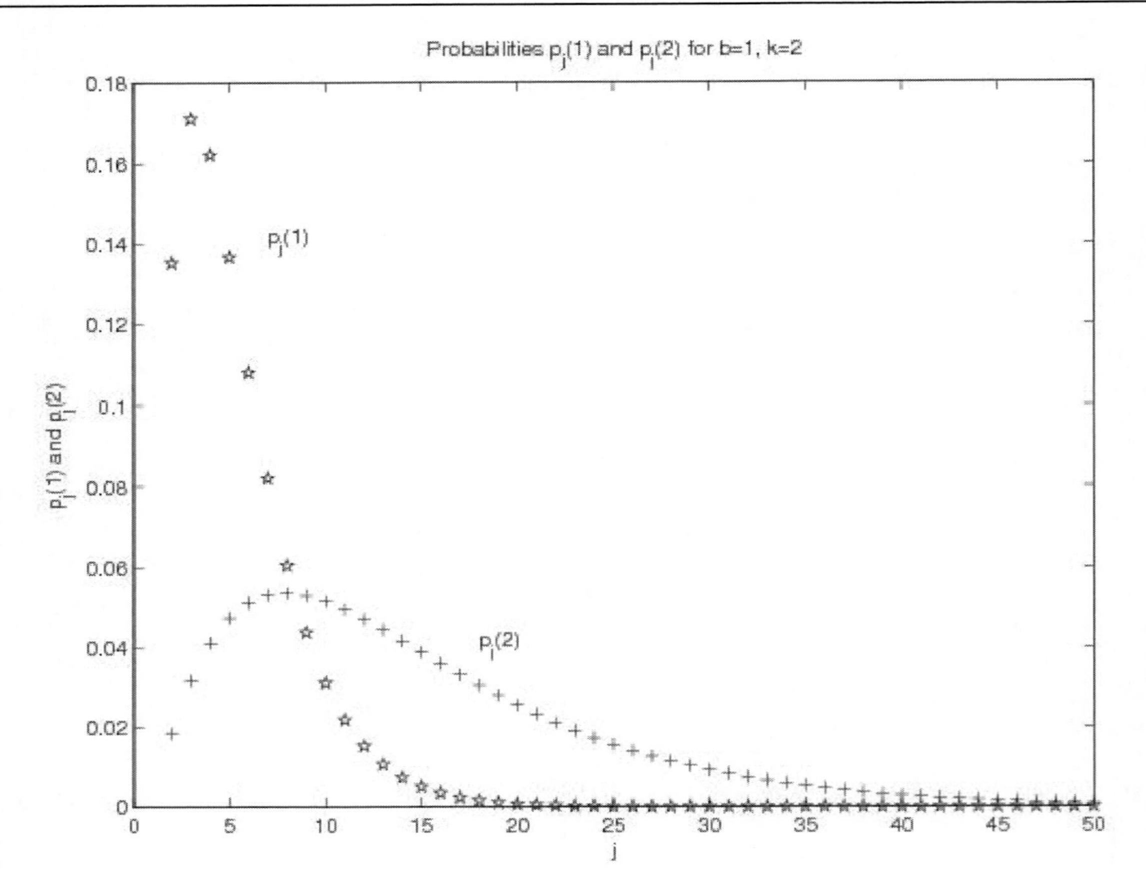

Figure 13.2. Probabilities $p_j(t)$ at time $t = 2$ for $b = 1, k = 2$

$$
\begin{aligned}
p'(t) &= \sum_{j=k}^{\infty} p_j'(t) \\
&= \sum_{j=k}^{\infty}(-jp_j(t)) + \sum_{j=k}^{\infty}(j-1)p_{j-1}(t) \quad (\text{with } p_{k-1} \equiv 0) \\
&= 0\,.
\end{aligned}
$$

Therefore, $p(t) \equiv p(0) = 1$. Our reasoning here lacks justification, however, since, in general, one is not allowed to differentiate a limit relation term by term. For example, we have

$$\frac{1}{n}\sin(n^2 t) \to 0 \quad \text{as} \quad n \to \infty ,$$

and, formally differentiating w.r.t. t,

$$n\cos(n^2 t) \to 0 \quad \text{as} \quad n \to \infty ,$$

which is nonsense.

Instead of trying to justify our formal computation, we now prove (13.4) using the explicit formula

$$p_j(t) = a_j e^{-jbt}(e^{bt} - 1)^{j-k}, \quad a_j = \begin{pmatrix} j-1 \\ k-1 \end{pmatrix} .$$

Set

$$w = e^{-bt} \quad \text{for} \quad b > 0, \quad t \geq 0 ,$$

and

$$z = 1 - w .$$

We then have $0 < w \leq 1$ and

$$0 \leq z < 1 .$$

With these notations,

$$\begin{aligned} p_j(t) &= a_j w^j \Big(\frac{1}{w} - 1\Big)^{j-k} \\ &= a_j w^k w^{j-k} \Big(\frac{1}{w} - 1\Big)^{j-k} \\ &= a_j w^k (1-w)^{j-k} \\ &= a_j (1-z)^k z^{j-k} \end{aligned}$$

We then have to show that

$$\sum_{j \geq k} a_j (1-z)^k z^{j-k} = 1 \quad \text{for} \quad 0 \leq z < 1 ,$$

or, equivalently,

$$\sum_{n=0}^{\infty} a_{k+n} z^n = (1-z)^{-k} . \tag{13.5}$$

Here

$$\begin{aligned} a_{k+n} &= \binom{k+n-1}{k-1} \\ &= \frac{(k+n-1)!}{(k-1)!n!} \\ &= \frac{1}{n!}\, k(k+1)\cdots(k+n-1)\ . \end{aligned}$$

The function

$$f(z) = (1-z)^{-k}$$

is analytic for $|z| < 1$ and, by calculus,

$$f^{(n)}(0) = k(k+1)\cdots(k+n-1)\ .$$

Therefore, the relation

$$\sum_{n=0}^{\infty} \frac{1}{n!}\, k(k+1)\cdots(k+n-1)z^n = (1-z)^{-k}\ , \tag{13.6}$$

which is equivalent to (13.5), is the Taylor series representation of the function $f(z) = (1-z)^{-k}$ about $z = 0$. In a course on complex variables, it is shown that (13.6) holds for all complex numbers z with $|z| < 1$. This completes the proof of equation (13.4).

5. The Expected Value of X_t

We first introduce the important concept *expected value* or *mean* of a general discrete random variable X. Then we show that X_t, the random variable modeling the stochastic growth process, has the expected value $E(X_t) = ke^{bt}$. This may not come as a surprise since $u(t) = ke^{bt}$ is precisely the solution of the corresponding deterministic model, $u' = bu, u(0) = k$.

5.1. The Expected Value of a Discrete Random Variable

A real valued random variable X is called discrete if its values α_j form a finite set or an infinite sequence. If the value α_j is attained with probability p_j, i.e.,

$$p_j = prob\Big(X = \alpha_j\Big), \quad j = 1, 2, \ldots$$

then

$$E(X) = \sum_j \alpha_j p_j \tag{13.7}$$

is called the *expected value* or *mean* of X. Here the sum is finite if X assumes only finitely many values α_j or has infinitely many terms if the values α_j form an infinite sequence. In the latter case, we assume that the series (13.7) converges. Thus, the expected value $E(X)$ is a weighted average of all possible values α_j of X. The probabilities $p_j = prob(X = \alpha_j)$ are the weights.

5.2. The Expected Value $E(X_t)$

Let us compute the expected value $E(X_t)$ where X_t is the random variable introduced in Section 2.. Here the time $0 \le t < \infty$ is arbitrary, but fixed. By solving the Forward Kolmogorov Equations, we have seen that X_t attains the value j with probability $p_j(t)$ where

$$p_j(t) = \begin{cases} 0 & \text{for} \quad 1 \le j < k \\ a_j e^{-jbt}(e^{bt} - 1)^{j-k} & \text{for} \quad j \ge k \end{cases} \tag{13.8}$$

and a_j is the binomial coefficient

$$a_j = \begin{pmatrix} j-1 \\ k-1 \end{pmatrix} .$$

The positive integer $k = X_0$ is the size of the population at time $t = 0$.

Thus, by definition, the expected value of X_t is

$$E(X_t) = \sum_{j \ge k} j p_j(t) .$$

We will now use the notations introduced in the previous section,

$$0 < w = e^{-bt} \le 1, \quad 0 \le z = 1 - w < 1 ,$$

and compute the value of the series defining $E(X_t)$. We have

$$\begin{aligned}
E(X_t) &= \sum_{j\geq k} j a_j e^{-jbt}(e^{bt}-1)^{j-k} \\
&= \sum_{j\geq k} j a_j w^j \Big(\frac{1}{w}-1\Big)^{j-k} \\
&= \sum_{j\geq k} j a_j w^k (1-w)^{j-k} \\
&= \sum_{j\geq k} j a_j (1-z)^k z^{j-k}
\end{aligned}$$

We claim that the value of the series equals ke^{bt}, i.e.,

$$\sum_{j\geq k} j a_j (1-z)^k z^{j-k} = kw^{-1} = k(1-z)^{-1} ,$$

or, equivalently

$$\sum_{j\geq k} j a_j z^{j-k} = k(1-z)^{-k-1} .$$

Setting $j-k=l$, we must show that

$$\sum_{l=0}^{\infty} (k+l) a_{k+l} z^l = k(1-z)^{-k-1} \quad \text{for} \quad 0 \leq z < 1 . \tag{13.9}$$

To show this, we recall from the previous section (see (13.5)) the equation

$$\sum_{n=0}^{\infty} a_{k+n} z^n = (1-z)^{-k} =: f(z) \quad \text{for} \quad |z| < 1 . \tag{13.10}$$

Since (13.10) is the power series representation of the analytic function $f(z) = (1-z)^{-k}$, we may differentiate the series term by term and obtain

$$\begin{aligned}
k(1-z)^{-k-1} &= \sum_{n=1}^{\infty} n a_{k+n} z^{n-1} \quad (\text{set } n-1 = l) \\
&= \sum_{l=0}^{\infty} (l+1) a_{k+l+1} z^l .
\end{aligned}$$

To arrive at (13.9), we must show that

$$(l+1)a_{k+l+1} = (k+l)a_{k+l} \ . \tag{13.11}$$

With $k+l+1=j$ this becomes

$$(j-k)a_j = (j-1)a_{j-1} \ ,$$

This is precisely the recursion relation (13.3) for the coefficients a_j which we have derived above to solve Kolmogorov's equations.

We have shown the following result.

Theorem 13.1. *The expected value of the random variable X_t modeling the growth process is*

$$E(X_t) = ke^{bt}, \quad t \geq 0 \ .$$

For later reference, we also note that we have shown the following two equations

$$\sum_{n=0}^{\infty} a_{k+n} z^n = (1-z)^{-k} = f(z) \tag{13.12}$$

$$\sum_{n=0}^{\infty} (k+n) a_{k+n} z^n = k(1-z)^{-k-1} = f'(z) \tag{13.13}$$

for $|z|<1$. When we compute the variance of X_t in the next section, it will be useful to also have a formula for $f''(z)$, i.e., to differentiate (13.13) one more time. We have

$$\begin{aligned} f''(z) &= (k^2+k)(1-z)^{-k-2} \\ &= \sum_{n=1}^{\infty} (k+n) n a_{k+n} z^{n-1} \\ &= \sum_{l=0}^{\infty} (k+l+1)(l+1) a_{k+l+1} z^l \quad \text{(use (13.11))} \\ &= \sum_{l=0}^{\infty} (k+l+1)(k+l) a_{k+l} z^l \\ &= \sum_{n=0}^{\infty} (k+n+1)(k+n) a_{k+n} z^n \end{aligned}$$

The equations for $f''(z)$ and $f'(z)$ yield

$$f''(z) - f'(z) = \sum_{n=0}^{\infty} (k+n)^2 a_{k+n} z^n \quad \text{for} \quad |z| < 1 \ . \tag{13.14}$$

6. The Variance of X_t

Consider two random variables, X and Y, and assume that X attains the values $9, 10, 11$, each with probability $\frac{1}{3}$. Then the mean of X is

$$E(X) = \frac{1}{3}(9 + 10 + 11) = 10 \ .$$

The second random variable Y attains the values $0, 10, 20$, each with probability $\frac{1}{3}$. The mean of Y is

$$E(Y) = \frac{1}{3}(0 + 10 + 20) = 10 \ .$$

Both random variables have the same mean, but X varies much less about the mean than Y. The variance of a random variable is a quantitative measure for the variation about its mean. Roughly speaking, the smaller the variance, the less variation occurs about the mean.

A precise definition of the variance $Var(X)$ of a general discrete random variable X will be given below. We then compute the variance of X_t and show that

$$Var(X_t) = ke^{bt}(e^{bt} - 1) \quad \text{for} \quad t \geq 0 \ .$$

6.1. The Variance of a Discrete Random Variable

Let us assume that X is a discrete random variable which attains the value α_j with probability

$$p_j = prob\Big(X = \alpha_j\Big) \quad \text{for} \quad j = 1, 2, \ldots \quad \text{where} \quad \sum_j p_j = 1 \ .$$

We recall from the previous section that

$$\mu := E(X) = \sum_j \alpha_j p_j$$

is the mean (or expected value) of X. Here the sum is finite or, if the values α_j form an infinite sequence, the series is assumed to converge absolutely. The same assumption holds for all the series below.

If we set $Y = X - \mu$ then the random variable Y attains the value $\alpha_j - \mu$ with probability p_j and, therefore, the mean of Y is

$$\begin{aligned} E(Y) &= \sum_j (\alpha_j - \mu) p_j \\ &= \sum_j \alpha_j p_j - \mu \sum_j p_j \\ &= \mu - \mu \\ &= 0 \end{aligned}$$

Definition 13.1: The number

$$Var(X) = E\Big((X-\mu)^2\Big) \quad \text{where} \quad \mu = E(X) \tag{13.15}$$

is called the variance of the random variable X. The number $\sigma = \sqrt{Var(X)}$ is called the standard deviation of X.

Since the random variable $Z = (X-\mu)^2$ attains the value $(\alpha_j - \mu)^2$ with probability p_j, we have

$$\begin{aligned} Var(X) &= E(Z) \\ &= \sum_j (\alpha_j - \mu)^2 p_j \\ &= \sum_j \alpha_j^2 p_j - 2\mu \sum_j \alpha_j p_j + \mu^2 \\ &= E(X^2) - \mu^2 \end{aligned}$$

This leads to the following useful formula

$$Var(X) = E(X^2) - \Big(E(X)\Big)^2 . \tag{13.16}$$

6.2. The Variance $Var(X_t)$

Recall that the random variable X_t, which models the stochastic growth process, attains the value j with probability $p_j(t)$ where $p_j(t)$ is given in (13.8). We want to show that $Var(X_t) = ke^{bt}(e^{bt} - 1)$. Using (13.16) and $E(X_t) = ke^{bt}$ we have

$$\begin{aligned} Var(X_t) &= E((X_t)^2) - k^2 e^{2bt} \\ &= \sum_{j \geq k} j^2 p_j(t) - k^2 e^{2bt} \end{aligned}$$

and we must show that

$$\sum_{j\geq k} j^2 p_j(t) = k^2 e^{2bt} + k e^{bt}(e^{bt} - 1) \ .$$

As above, we will use the variables w and z where

$$0 < w = e^{-bt} \leq 1 \quad \text{and} \quad 0 \leq z = 1 - w < 1 \ .$$

The above claim then becomes

$$\sum_{j\geq k} j^2 a_j w^j \Big(\frac{1}{w} - 1\Big)^{j-k} = k^2 w^{-2} + k w^{-1}(w^{-1} - 1)$$

or

$$\sum_{j\geq k} j^2 a_j (1-w)^{j-k} = (k^2 + k) w^{-k-2} - k w^{-k-1}$$

or (write $w^j = w^{j-k} w^k$ and divide by w^k)

$$\sum_{j\geq k} j^2 a_j z^{j-k} = (k^2 + k)(1-z)^{-k-2} - k(1-z)^{-k-1} =: rhs \ . \tag{13.17}$$

If $f(z) = (1-z)^{-k}$ then the right–hand side is $f''(z) - f'(z)$ and, using (13.14), we have

$$\begin{aligned} rhs &= f''(z) - f'(z) \\ &= \sum_{n=0}^{\infty} (k+n)^2 a_{k+n} z^n \\ &= \sum_{j\geq k} j^2 a_j z^{j-k} \end{aligned}$$

which proves (13.17).

We summarize our results about the random variable X_t.

Theorem 13.2. *Let X_t denote the random variable modeling the stochastic growth process described in Section 2.. The probabilities*

$$p_j(t) = prob\Big(X_t = j\Big)$$

satisfy the Forward Kolmogorov Equations (13.1). The expected value and variance of X_t are

$$\begin{aligned} E(X_t) &= k e^{bt} \ , \\ Var(X_t) &= k e^{bt}(e^{bt} - 1) \ . \end{aligned}$$

Here $X_0 = k$ is the size of the population at time $t = 0$.

7. Statistics of Interevent Times

The random variable X_t describing the growth process takes values in the set of integers $\{k, k+1, k+2, \ldots\}$ where $X_0 = k$ is the number of individuals in the population at time $t = 0$. Any realization of X_t will be determined by a sequence of times

$$T_0 = 0 < T_1 < T_2 < T_3 < \ldots$$

with a birth occurring at $T_1, T_2, \ldots$. A typical graph for X_t is given in Figure 13.3.

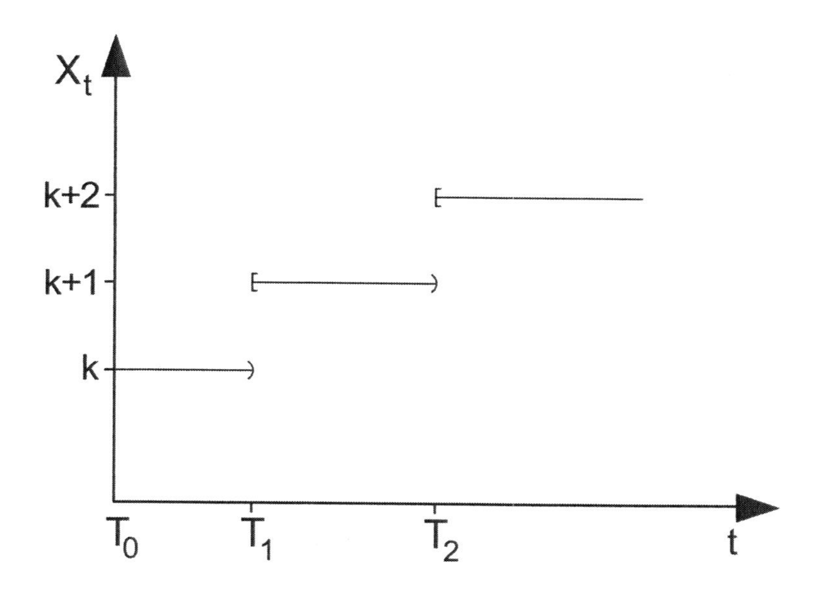

Figure 13.3. The random variable X_t with births at T_1 and T_2

A jump from $X_t = k + j$ to $X_t = k + j + 1$ occurs as time t increases from $T_j - \varepsilon$ to T_j. But what can we say about the distribution of the random times T_j at which a birth occurs?

A more precise mathematical question is: What is the statistical law for the T_j that is consistent with the Forward Kolmogorov Equations (13.1) for the random variable X_t?

The positive times

$$\tau_j = T_{j+1} - T_j, \quad j = 0, 1, \ldots$$

are called interevent times. For example, $\tau_5 = T_6 - T_5$ is the time between the occurrence of the birth (an event) at time T_5 and the occurrence of the next birth at time T_6. Going back to our original description of the growth process, we will derive the following statistical law for τ_j:

$$prob\Big(\tau_j > \tau\Big) = e^{-b(k+j)\tau} \quad \text{for all} \quad \tau \geq 0 \ . \tag{13.18}$$

Let us try to understand the meaning of the formula and assume, for example, that

$$b = \frac{1}{1\,hour}, \quad k = 1, \quad j = 2 \ .$$

The population started with $k = 1$ individual (cell) and two births (cell divisions) have occurred, one at time T_1 and the next at time T_2. The formula (13.18) gives information about the time $T_3 = T_2 + \tau_2$ at which one can expect the next birth to occur. For example, since

$$e^{-3} = 0.0498\ldots$$

the probability for τ_2 to exceed one hour is

$$prob(\tau_2 > 1\,hour) = e^{-3} = 0.0498\ldots$$

The probability for the next birth (cell division) after T_2 to occur within one hour is $1-e^{-3}$, corresponding to about 95%.

In Section 8. we will show how the formula (13.18) can be used to run numerical realizations for the random variable X_t.

To derive the formula (13.18) we will need a basic result on *conditional probability.* Let us explain the result in general.

7.1. Conditional Probability

Suppose E_1 and E_2 are two events with probabilities $prob(E_1)$ and $prob(E_2)$, respectively. What can be said about the probability that both events, E_1 and E_2, will occur? We want to explain the formula

$$prob\Big(E_1 \text{ and } E_2\Big) = prob\Big(E_1\Big) \cdot prob\Big(E_2|E_1\Big) \tag{13.19}$$

where $prob(E_2|E_1)$ is the probability of the occurrence of E_2 *under the assumption of the occurrence of* E_1. One calls $prob(E_2|E_1)$ a conditional probability.

We want to illustrate (13.19) by a concrete example. Suppose two fair dice are cast, die one shows A and die two shows B,

$$A, B \in \{1, 2, 3, 4, 5, 6\} \ .$$

There are 36 possible outcomes for the pair (A, B) and each pair occurs with probability $\frac{1}{36}$. Consider the following two events:

E_1 is the event $A + B \leq 3$ and E_2 is the event $A = 1$.

Clearly, $prob(E_2) = \frac{1}{6}$. The following three outcomes for (A, B) lead to E_1:

$$(1, 1), \quad (1, 2), \quad (2, 1) \ . \tag{13.20}$$

Therefore, $prob(E_1) = \frac{3}{36} = \frac{1}{12}$. What is the probability that both events, E_1 and E_2, occur? Only the two pairs

$$(1, 1), \quad (1, 2)$$

correspond to the occurrence of E_1 and E_2, thus $prob(E_1 \text{ and } E_2) = \frac{2}{36} = \frac{1}{18}$.

If we assume that E_1 occurs, then the only possible outcomes for (A, B) are listed in (13.20). In two of the three outcomes we have $A = 1$. Therefore, $prob(E_2|E_1) = \frac{2}{3}$. Altogether,

$$\begin{aligned} prob\Big(E_1\Big) \cdot prob\Big(E_2|E_1\Big) &= \frac{1}{12} \cdot \frac{2}{3} \\ &= \frac{1}{18} \\ &= prob\Big(E_1 \text{ and } E_2\Big) \end{aligned}$$

which confirms the formula (13.19).

Remark: In an abstract approach to probability theory due to Kolmogorov, equation (13.19) is used to *define* $prob(E_2|E_1)$. Equation (13.19) then becomes a triviality, true by definition of $prob(E_2|E_1)$. The abstract approach to probability theory is very useful if the main emphasis is to obtain a *mathematically consistent* theory.

7.2. Statistical Law for Interevent Times

In this section, we want to derive the formula

$$prob(\tau_j > \tau) = e^{-b(k+j)\tau} \quad \text{for} \quad \tau \geq 0 \tag{13.21}$$

for the random variable $\tau_j = T_{j+1} - T_j$. To this end, we set

$$G_j(\tau) := prob(\tau_j > \tau)$$

and will use our original modeling assumptions to derive a differential equation for the function $G_j(\tau)$.

In the following, the time $\tau \geq 0$ and the (small) time increment $\Delta\tau$ are considered to be fixed. We introduce two events and their probabilities:

E_1 is the event: $\tau_j > \tau$.

E_2 is the event: No birth occurs in the time interval $[T_j + \tau, T_j + \tau + \Delta\tau]$.

By definition, we have

$$\begin{aligned} G_j(\tau) &= prob(\tau_j > \tau) = prob(E_1) \\ G_j(\tau + \Delta\tau) &= prob(\tau_j > \tau + \Delta\tau) \end{aligned}$$

If E_1 occurs then $X_{T_j+\tau} = k + j$ and if — additionally — E_2 occurs, then there is no birth in the time interval $[T_j + \tau, T_j + \tau + \Delta\tau]$. Therefore, E_1 and E_2 imply that

$$X_{T_j+\tau+\Delta\tau} = k + j \; .$$

It is also clear that the converse holds: If $X_{T_j+\tau+\Delta\tau} = k + j$ then the events E_1 and E_2 have occurred. But the equality $X_{T_j+\tau+\Delta\tau} = k + j$ is equivalent to $\tau_j > \tau + \Delta\tau$. We therefore obtain that

$$prob(\tau_j > \tau + \Delta\tau) = prob(E_1 \text{ and } E_2) \; .$$

To summarize,

$$\begin{aligned} G_j(\tau + \Delta\tau) &= prob(\tau_j > \tau + \Delta\tau) \\ &= prob(E_1 \text{ and } E_2) \\ &= prob(E_1) \cdot prob(E_2|E_1) \\ &= G_j(\tau) \cdot prob(E_2|E_1) \end{aligned}$$

We now determine the conditional probability $prob(E_2|E_1)$ using our original modeling assumptions. Recall that the probability for an individual to give birth in a time interval of length $\Delta\tau$ is $\sim b\Delta\tau$ for small $\Delta\tau$. Therefore, if there are $k + j$ individuals, then the probability that *no* birth occurs in a time interval of length $\Delta\tau$ is

$$\sim 1 - (k + j)b\Delta\tau \; .$$

Assuming E_1, there are $k + j$ individuals at time $T_j + \tau$. Therefore, the conditional probability $prob(E_2|E_1)$ is

$$prob(E_2|E_1) \sim 1 - (k + j)b\Delta\tau \; .$$

Using this result, we finally obtain

$$G_j(\tau + \Delta\tau) \sim G_j(\tau)\Big(1 - (k + j)b\Delta\tau\Big)$$

or

$$\frac{1}{\Delta\tau}\Big(G_j(\tau + \Delta\tau) - G_j(\tau)\Big) \sim -b(k + j)G_j(\tau) \; .$$

As $\tau \to 0$ one obtains the differential equation

$$G_j'(\tau) = -b(k+j)G_j(\tau) \ .$$

The initial condition

$$G_j(0) = prob(\tau_j > 0) = 1$$

then leads to the formula (13.21).

8. Numerical Realization of Random Evolution

We have seen in the previous section that the interevent time

$$\tau_j = T_{j+1} - T_j$$

(the time between the j–th and the $(j+1)$–th birth) is a random variable which satisfies the statistical law

$$prob\Big(\tau_j > t\Big) = e^{-b(k+j)t} \quad \text{for all} \quad 0 \le t < \infty \ . \tag{13.22}$$

How can we use this to run numerical realizations for X_t? Clearly, if the coefficient

$$b(k+j) > 0$$

is given, we need a method which gives as random numbers τ_j satisfying (13.22). We will now show how this can be done using Matlab's random number generator and a simple transformation.

The Matlab command

$$Y = rand$$

gives us random numbers Y with $0 < Y < 1$ which are uniformly distributed in the unit interval, i.e.,

$$prob(Y < Y_0) = Y_0 \quad \text{for all} \quad 0 < Y_0 < 1 \ . \tag{13.23}$$

Now let

$$F : [0, \infty) \to (0, 1]$$

denote any continuous, strictly decreasing function with

$$F(0) = 1, \quad \lim_{t \to \infty} F(t) = 0 \ .$$

The function

$$F(t) = e^{-b(k+j)t}$$

is an example. We then claim that the random variable [1]

$$T = F^{-1}(Y)$$

satisfies

$$prob(T > t) = F(t) \quad \text{for all} \quad 0 < t < \infty$$

if Y is uniformly distributed in $0 < Y < 1$. The reason is simple: The condition

$$T = F^{-1}(Y) > t$$

is equivalent to

$$Y < F(t) .$$

Therefore,

$$\begin{aligned} prob(T > t) &= prob(Y < F(t)) \\ &= F(t) \end{aligned}$$

The second equation is nothing but (13.23).

Clearly, we can apply the result to the function

$$F(t) = e^{-b(k+j)t}, \quad 0 \le t < \infty .$$

For given $0 < Y \le 1$, we determine $\tau_j = F^{-1}(Y)$ by solving the equation

$$Y = e^{-b(k+j)\tau_j}$$

for τ_j, which yields

$$\tau_j = \frac{\ln(1/Y)}{b(k+j)} .$$

If the random variable $0 < Y \le 1$ is uniformly distributed in the unit interval, then the numbers τ_j satisfies the law (13.22).

Once one has values for the interevent times τ_j, it is clear how the random variable X_t evolves: If one sets

[1] Here $F^{-1} : (0,1] \to [0,\infty)$ denotes the inverse function of $F : [0,\infty) \to (0,1]$; thus $F(F^{-1}(Y)) = Y$ for all $0 < Y \le 1$.

$$T_0 = 0, \quad T_{j+1} = T_j + \tau_j \quad \text{for} \quad j = 0, 1, \ldots$$

then

$$X_t = k + j \quad \text{for} \quad T_j \leq t < T_{j+1} \; .$$

9. Figures and Scripts

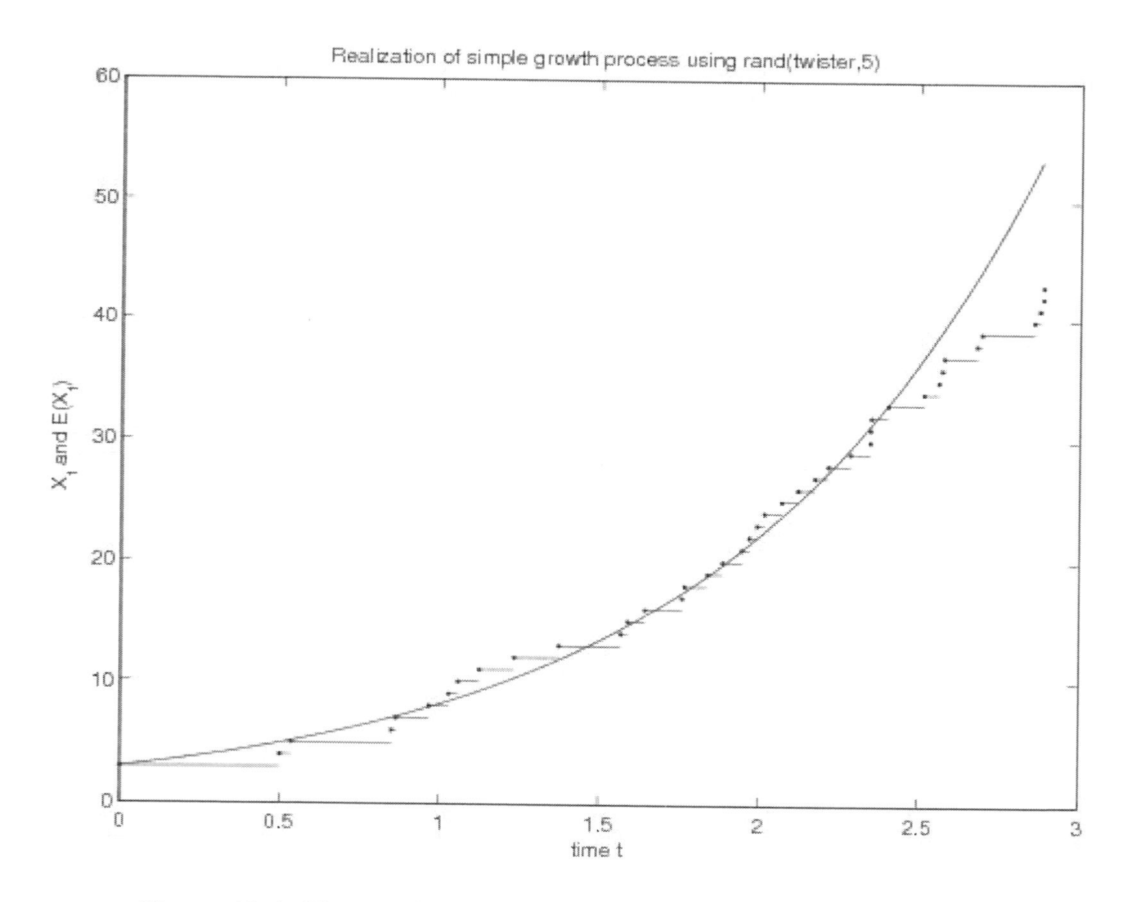

Figure 13.4. First realization of random evolution for $b = 1, k = 3$

In Figure 13.4 we show a numerical realizations of the evolution of X_t for 40 birth events assuming $b = 1, k = 3$. The figure also shows the expected value $E(X_t) = 3e^t$. The figure is obtained with the following Matlab script:

```
%b.m
k=3;
maxstep=40;
T=zeros(1,maxstep+1);
rand('twister',5);
for j=1:maxstep
y=rand;
T(j+1)=T(j)+log(1/y)/(k-1+j);
end
plot(T,[k:maxstep+k],'.'); hold on
newt=[T(1:maxstep -1);T(2:maxstep)];
for j=1:maxstep -1
plot(newt(:,j),[k+j-1 k+j-1],'-'); hold on
end
y=[0:0.01:T(end)];
z=k*exp(y);
plot(y,z,'-');
title('Realization of simple growth process using rand(twister,5)');
xlabel('time t');
ylabel('X_t and E(X_t)');
```

A comment on the fifth line of the script
rand('twister',5)
is in order: Matlab provides different methods to produce pseudo–random numbers. One of the methods is called the Mersenne Twister algorithm by Makoto Matsumoto and Takuji Nishimura [13]. The value 5 in the above line initializes the algorithm. If one uses 6 instead of 5, then Matlab chooses different random numbers and one obtains Figure 13.5. To obtain results that can be reproduced one should always initialize the random number generator.

Recall the definition

$$p_j(t) := prob\Big(X_t = j\Big) \; ,$$

i.e., $p_j(t)$ is the probability that the random variable X_t attains the value j at time t. Assuming that $X_0 = k$, we have derived the formula

$$p_j(t) = \begin{pmatrix} j-1 \\ k-1 \end{pmatrix} e^{-jbt}(e^{bt}-1)^{j-k}, \quad j \geq k \tag{13.24}$$

by solving the Forward Kolmogorov Equations in Section 3.. We can now use numerical simulations to check the formula.

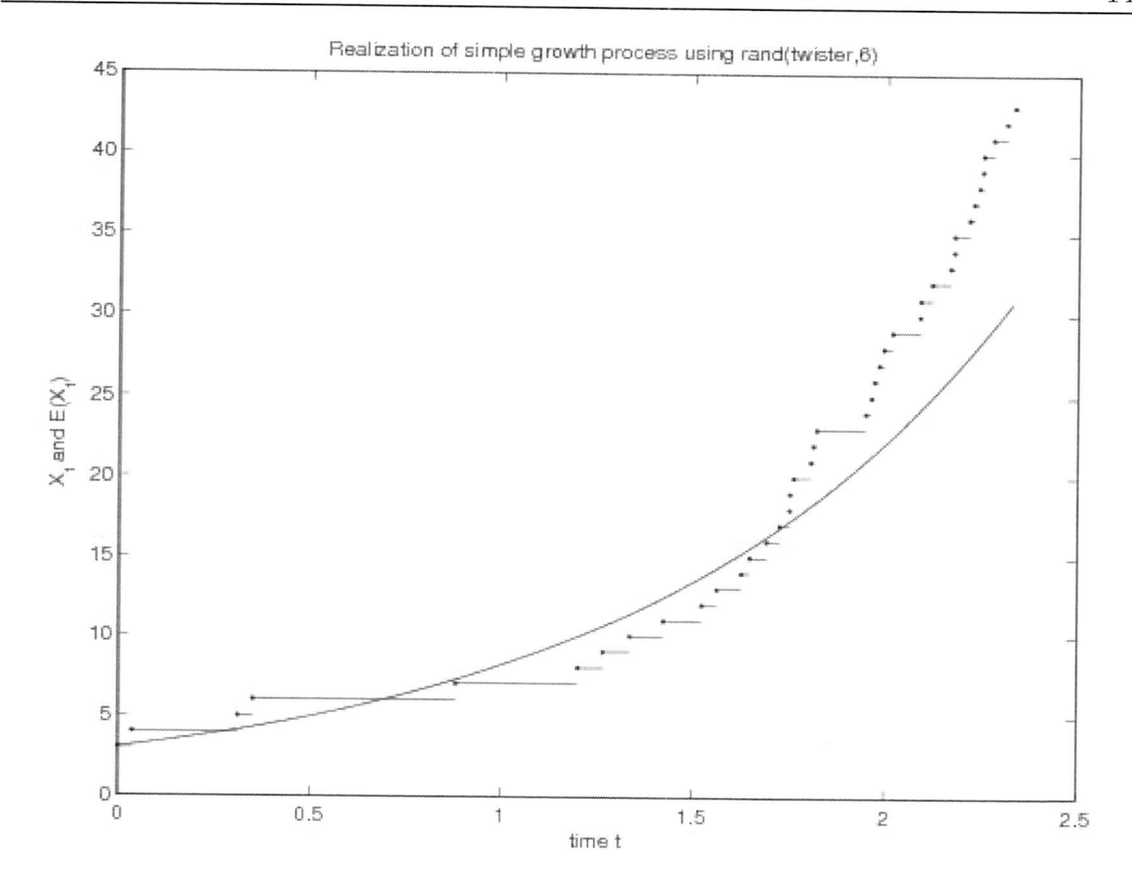

Figure 13.5. Second realization of random evolution for $b = 1, k = 3$

We set $b = 1, t = 3$, fix $X_0 = k \in \{1, 2, 3, 4\}$, and run 90,000 numerical simulations to obtain approximations for $p_j(3)$.

The results obtained with the following script are shown in Figure 13.6. The figure also shows four curves, which are obtained by plotting the probabilities $p_j(3)$ based on formula (13.24).

```
%%%%%%%%%%%%%%%%%%%%%%%%%%%%%%%%%%%%%%%%%%%%%%%%%%%%%%
%
% This script plots probabilities p_j(t) for t=3
% calculated numerically via 90000 experiments
% together with p_j(t) obtained analytically.
%
%%%%%%%%%%%%%%%%%%%%%%%%%%%%%%%%%%%%%%%%%%%%%%%%%%%%%%
```

```
% defining number of experiments
exper=90000;

% loop for different initial conditions
for x0=1:4
    maxstep=150;

    res=zeros(1,exper);

    for l=1:exper
        tau=0;
        for i=1:maxstep
            y=rand;
            tau=tau+log(1/y)/(x0-1+i);

            if tau>=3
                res(l)=i;
                break;
            end;
        end

    end;

    notinform=length(find(res==0));
    probab=zeros(1,max(res));

    analytprobab=zeros(max(res),1);

    for l=1:max(res)
        probab(l)=length(find(res==l));
        if l>=x0
            % This part could be done more efficiently
            analytprobab(l)=factorial(l)/factorial(x0-1)/...
                factorial(l+1-x0)*exp(-3*l)*(exp(3)-1)^(l-x0);
        end;
    end;

    plot([1:max(res)],analytprobab,'-');hold on;
    plot([1:max(res)],probab/(exper-notinform),'.');
```

```
end;
title('p_j(3) for X_0 from 1 to 4');
xlabel('j');
ylabel('p_j(3)');
```

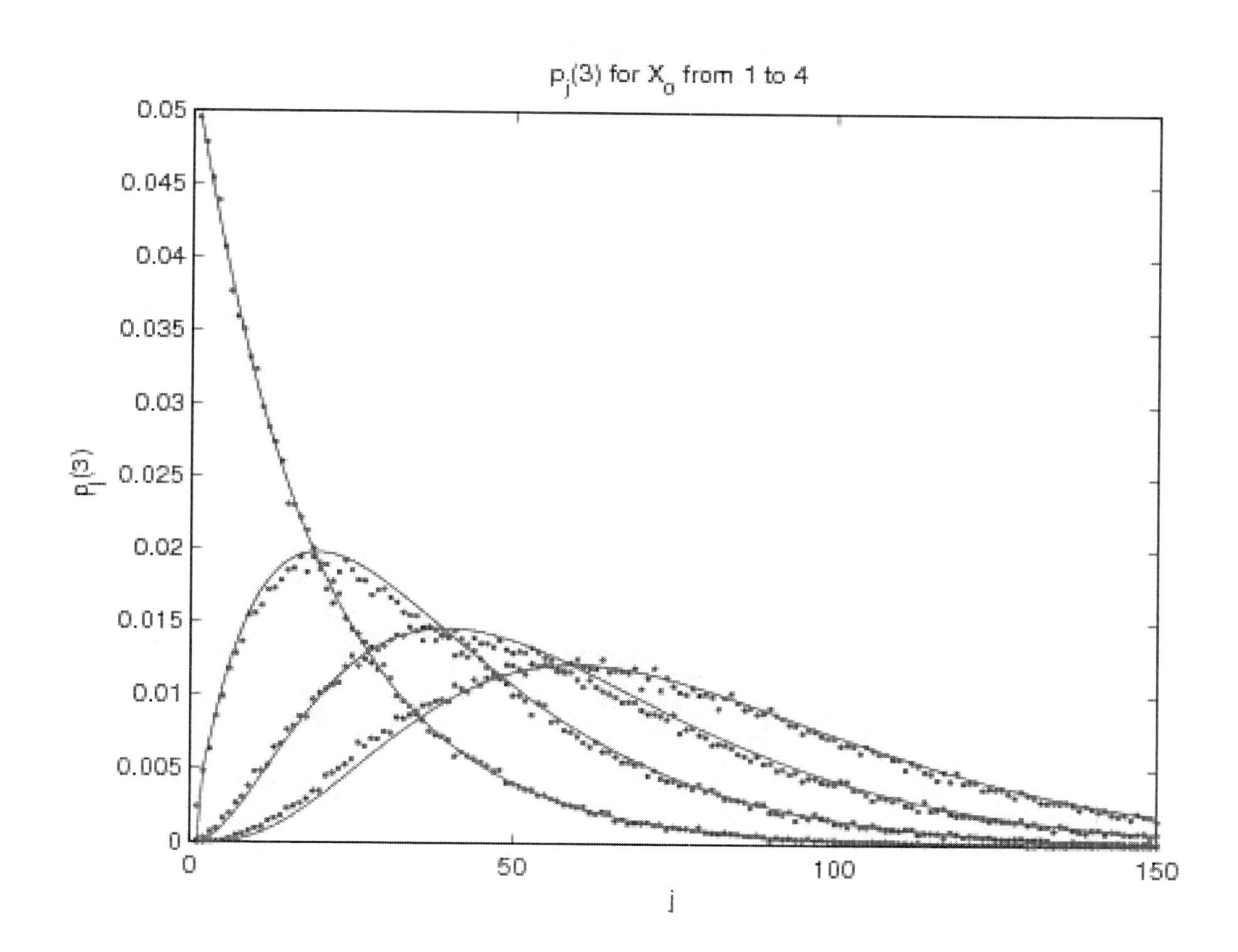

Figure 13.6. Probabilities $p_j(3)$ obtained via 90,000 experiments and analytically

Chapter 14

Introduction to Kinetic Theory

Summary: Two grams of hydrogen gas consist of about $N_A = 6.02 \cdot 10^{23}$ H_2–molecules. The number N_A, called Avogadro's number, is enormously large. If every person on earth had ten trillion dollars, the total number of dollars would be $7 \cdot 10^9 \cdot 10^{13} = 7 \cdot 10^{22}$, still short of Avogadro's number. The pure size of N_A makes it practically impossible to describe the motion of all the particles in a (macro amount) of a gas *individually*. Only a statistical description is possible.

In this chapter we give an introduction to kinetic theory. The basic aim is to connect the micro models for fluids and gases (bouncing particles) with macro models, as described by thermodynamics and continuum mechanics. We can also think of relating random evolution on the microscale to deterministic evolution on the macroscale.

A first result of kinetic theory dates back to 1738 when Daniel Bernoulli explained Boyle's law ($pV = const$) using a bouncing ball model. We give Bernoulli's arguments in Section 2.. Maxwell's velocity distribution is the subject of Section 3., and in Section 4. we describe Sadi Carnot's heat engine.

1. Boyle, Bernoulli, Maxwell, and Sadi Carnot

In 1662, the British scientist Robert Boyle (1627–1691) published the gas law

$$pV = const \quad \text{(at constant temperature)} \tag{14.1}$$

now known as Boyle's law. (The French call it Mariotte's law.) Suppose a gas occupies a volume V at pressure p. Then, if the volume is changed to V_1 and the temperature remains unchanged, Boyle's law says that the pressure changes to p_1 where

$$pV = p_1 V_1 \ .$$

How can one understand this empirical law if one thinks of the gas as a large number of particles bouncing around in the occupied volume? Daniel Bernoulli (1700–1782) addressed this question in 1738 and obtained the first remarkable result of kinetic theory, a micro model explanation of Boyle's law. We will follow Daniel Bernoulli in Section 2..

When deriving Boyle's law, Bernoulli assumed, wrongly, that all particles in the gas have the same speed. James Clerk Maxwell (1831–1879) was the first to introduce statistical concepts into physics. In 1860, he derived a distribution function for the speeds of particles in a gas. We present Maxwell's distribution in Section 3.. It turns out that Maxwell's distribution function contains a free parameter which can be related to the absolute temperature T of the gas. Using the relation

$$\frac{m}{2}\langle v^2 \rangle = \frac{3}{2} kT \ ,$$

where $\langle v^2 \rangle$ is the mean of v^2 and k is Boltzmann's constant, the free parameter in Maxwell's distribution can be expressed in terms of the absolute temperature T. This results in the Maxwell–Boltzmann distribution.

In the second half of the 17 hundreds, the Scottish engineer James Watt [1] (1736–1819) radically improved the efficiency of steam engines. His invention was very important for the Industrial Revolution. Watt's engineering work also led to fundamental questions. Somewhat vaguely, one can ask: Given an amount Q_1 of heat energy at temperature T_1, how much of the heat energy Q_1 can be turned into (useful) mechanical work? Formulated in this way, the question really makes no sense. Any heat engine, which turns heat energy into mechanical energy, works between (at least) two temperatures. Thus, a more meaningful question is: Given an amount of heat energy Q_1 at temperature T_1 and assuming a temperature T_2 less than T_1 of the surroundings, how much of the heat energy Q_1 can be turned into mechanical energy? If a heat engine produces the mechanical energy E_1 using the heat energy Q_1, one says that its efficiency is

$$\eta = E_1/Q_1 \ .$$

Roughly, the efficiency is defined as the quotient

$$\frac{\text{what one gets out}}{\text{what one puts in}} \ .$$

An efficiency $\eta = 1$ corresponds to a heat engine which turns the heat energy Q_1 completely into mechanical work. Does such a heat engine exist, at least theoretically? Or is the optimal efficiency η always strictly less than one? How does η depend on the temperatures T_1 and T_2?

Such questions were addressed, with most remarkable insights, by the French engineer Sadi Carnot (1796–1832), at a time when neither the nature of heat energy nor the principle of conservation of energy were well understood. Carnot described an idealized heat engine with *optimal efficiency*. His work had a major impact on the development

[1] The unit of power, $1watt = 1Joule/second$, is named after him.

of thermodynamics. The roots of the concept of entropy, introduced by Rudolf Clausius (1822–1888) around 1850, lie in Carnot's work. It also inspired the later development of the diesel engine, whose high temperature gives it greater efficiency. We will describe Carnot's idealized heat engine, formalized as the Carnot cycle, in Section 4..

2. Daniel Bernoulli: Pressure, Volume, and Particle Velocity

Consider a box $B = [0, L]^3$ with volume $V = L^3$ which contains N particles, each of mass m. Let p denote the pressure of the gas and let T denote its temperature. [2]

Suppose we can change the volume of the box by moving a wall. This will change the temperature, T, but we wait until the temperature is again the temperature of the surroundings. Let p_1, V_1 denote pressure and volume after the wall has been moved.

One observes that (for an ideal gas)

$$pV = p_1 V_1 \ .$$

This law, $pV = const$ at constant temperature, is called Boyle's law. In trying to understand the law, D. Bernoulli obtained a first result of kinetic theory.

Suppose all the particles move at the same speed, v. Also, assume that at any given time, $\frac{N}{6}$ of the particles move in the positive x direction, i.e., have velocity vector

$$\mathbf{v} = (v, 0, 0) \ . \tag{14.2}$$

These simplifying assumptions are a strike of genius of Bernoulli. They are clearly wrong, but make the problem treatable.

Let W denote the wall parallel to the yz–plane at $x = L$ and consider a particle with velocity $\mathbf{v} = (v, 0, 0)$ before it hits the wall W. After it hits W, its velocity is $-\mathbf{v}$.

Newton's law,

$$m \frac{d}{dt} \mathbf{v}(t) = force \ ,$$

cannot be applied directly if we have an instantaneous change of velocity from $(v, 0, 0)$ to $(-v, 0, 0)$ because the required force would be infinite. However, we can argue as follows: Let

$$\mathbf{v}(t) = (v, 0, 0)$$

and

[2] With T, T_1, etc. we will always denote absolute temperatures although this concept was unknown in Bernoulli's time.

$$\mathbf{v}(t+\Delta t) = (-v, 0, 0) \ .$$

We then have

$$m\Big(\mathbf{v}(t+\Delta t) - \mathbf{v}(t)\Big) \sim force \cdot \Delta t$$

where the force is the force of the wall acting on the particle. Conversely, if $\mathbf{F} = (f, 0, 0)$ is the force of the particle applied to W, then

$$-m\Big(\mathbf{v}(t+\Delta t) - \mathbf{v}(t)\Big) \sim \mathbf{F} \cdot \Delta t \ .$$

This yields

$$2mv \sim f\Delta t \ .$$

(Recall that v and f are the magnitudes of the vectors $\mathbf{v}$ and $\mathbf{F}$.)

In a time interval of length Δt, the number of particles that will hit the wall W is

$$N_1 = \frac{N}{6} \cdot \frac{v\Delta t}{L} \ .$$

Thus, if F is the magnitude of the total force on the wall W, then

$$\frac{N}{6} \cdot \frac{v\Delta t}{L} \cdot 2mv \sim F\Delta t \ .$$

We divide by Δt and obtain

$$F = \frac{2N}{3} \cdot \frac{1}{2} mv^2 \cdot \frac{1}{L} \ .$$

The pressure on the wall W is

$$p = F/L^2 = \frac{2N}{3} \cdot \frac{1}{2} mv^2 \cdot \frac{1}{L^3} \ .$$

In other words,

$$pV = \frac{2N}{3} \cdot \frac{1}{2} mv^2 \ . \tag{14.3}$$

Here $\frac{1}{2}mv^2$ is the kinetic energy of any of the N particles. Equation (14.3) is D. Bernoulli's result. If one assumes that a constant temperature corresponds to a constant kinetic energy of each of the particles, then (14.3) explains Boyle's law (14.1).

2.1. The Ideal Gas Law

As it stands, Bernoulli's result (14.3) needs a modification: One has to replace v^2 in (14.3) by its mean. A correct result is:

$$pV = \frac{2N}{3} \cdot \frac{m}{2}\langle v^2 \rangle \tag{14.4}$$

where $\langle v^2 \rangle$ is the mean of v^2 over all particle. (In fact, how to take the mean, is not completely clear.)

It was also discovered later that there is a universal constant k so that

$$\frac{3}{2}kT = \frac{m}{2}\langle v^2 \rangle \tag{14.5}$$

where T is the absolute temperature of the gas. The constant k is Boltzmann's constant[3]

$$k = 1.3804 \cdot 10^{-23}\, Joule/K \ .$$

Here K is one degree Kelvin and

$$1\, Joule = 1\, Newton\ meter = 1\, Watt\ sec \ .$$

(The unit of force $1 Newton = 1kg\ meter/sec^2$ is the force which accelerates a mass of one kilogram by one meter per second squared. More concretely, one Newton is about the weight of an apple.)

From (14.4) and (14.5) one obtains

$$\begin{aligned} pV &= NkT \\ &= \frac{N}{N_A}\, kN_A T \\ &= nRT \end{aligned}$$

Here (with Avogadro's constant $N_A = 6.02 * 10^{23}$)

$$n = \frac{N}{N_A}\, mol$$

and

$$R = \frac{kN_A}{mol} = 8.314\, \frac{Joule}{mol K}$$

[3]Boltzmann's constant describes the gap between macroscopic and microscopic physics. In formula (14.8) given below it is good to think of kT as an energy which is of the order of magnitude of the average energy of a particle in a substance of absolute temperature T.

is the universal gas constant. The number N/N_A is the number of moles of the gas; by definition, one mole of a substance contains N_A molecules. The resulting formula

$$pV = nRT \tag{14.6}$$

is an example of an equation of state, called the universal gas law. We will use it in Section 4. on Carnot's cycle.

3. The Maxwell–Boltzmann Distribution

In 1860, James Clerk Maxwell derived a formula for the distribution of the speeds of particles in a gas. Remarkably, his derivation was solely based on reasonable symmetry assumptions. Maxwell's distribution function contains a free parameter, and if one uses the equation $\frac{m}{2}\langle v^2\rangle = \frac{3}{2}kT$, the free parameter can be expressed in terms of the absolute temperature, T. The result is the well–known Maxwell–Boltzmann distribution. We will derive it in this section.

First, some integrals:

Lemma 14.1.

$$\begin{aligned}
\int_0^\infty e^{-x^2}\,dx &= \frac{1}{2}\sqrt{\pi} \\
\int_0^\infty x^2 e^{-x^2}\,dx &= \frac{1}{4}\sqrt{\pi} \\
\int_0^\infty x^4 e^{-x^2}\,dx &= \frac{3}{2^3}\sqrt{\pi} \\
\int_0^\infty x^{2n} e^{-x^2}\,dx &= \frac{1\cdot 3\cdots(2n-1)}{2^{n+1}}\sqrt{\pi} \quad \textit{for} \quad n = 1, 2, \ldots
\end{aligned}$$

Proof: Let $J = \int_{-\infty}^{\infty} e^{-x^2}\,dx$. Then, using polar coordinates,

$$\begin{aligned}
J^2 &= \int_{\mathbb{R}^2} e^{-(x^2+y^2)}\,dxdy \\
&= \int_0^{2\pi}\int_0^\infty re^{-r^2}\,drd\phi \\
&= \pi\int_0^\infty 2re^{-r^2}\,dr \quad (\text{substitute } \xi = r^2) \\
&= \pi\int_0^\infty e^{-\xi}\,d\xi \\
&= \pi
\end{aligned}$$

This proves the first formula. To prove the second formula, we integrate by parts:

$$\begin{aligned}
\int_0^\infty x^2 e^{-x^2}\,dx &= \int_0^\infty \frac{x}{2}\,(2xe^{-x^2})\,dx \\
&= \int_0^\infty \frac{x}{2}(-e^{-x^2})'\,dx \\
&= \int_0^\infty \frac{1}{2}\,e^{-x^2}\,dx \\
&= \frac{1}{4}\sqrt{\pi}
\end{aligned}$$

The 3rd formula follows similarly,

$$\begin{aligned}
\int_0^\infty x^4 e^{-x^2}\,dx &= \int_0^\infty \frac{x^3}{2}\,(2xe^{-x^2})\,dx \\
&= \int_0^\infty \frac{x^3}{2}\,(-e^{-x^2})'\,dx \\
&= \int_0^\infty \frac{3}{2}\,x^2 e^{-x^2}\,dx \\
&= \frac{3}{2^3}\sqrt{\pi}
\end{aligned}$$

The general case follows by induction in n. $\diamond$

Consider a gas in thermodynamic equilibrium consisting of a large number of identical particle, each of mass m. (In reality, each particle will be an atom or a molecule.) (We assume that we have chosen units for length and time so that the velocity vectors $\mathbf{v} \in \mathbb{R}^3$ of the particles are dimensionless.

If $B \subset \mathbb{R}^3$ is any measurable set and if $\mathbf{v} \in \mathbb{R}^3$ is the velocity vector of a randomly chosen particle of the gas, we can ask: What is the probability for *the event* $\mathbf{v} \in B$? The Maxwell–Boltzmann distribution function, which we will derive, gives an answer in terms of an integral:

$$prob(\mathbf{v} \in B) = \int_B f(\mathbf{v})d\mathbf{v} \tag{14.7}$$

with

$$f(\mathbf{v}) = \left(\frac{m}{2\pi kT}\right)^{3/2} e^{-mv^2/2kT} \quad \text{with} \quad v = |\mathbf{v}|\ . \tag{14.8}$$

Here k is Boltzmann's constant and T is the absolute temperature of the gas.

We now formalize five assumptions which will lead to Maxwell's distribution. Throughout, $B = I_1 \times I_2 \times I_3 \subset \mathbb{R}^3$ denotes a box in velocity space; the box B is the product of three finite intervals $I_j \subset \mathbb{R}$. Also,

$$\mathbf{v} = (v_1, v_2, v_3)$$

denotes the velocity vector of a randomly chosen particle.

3.1. Assumptions Leading to Maxwell's Distribution

1. There is a smooth function $f : \mathbb{R}^3 \to (0, \infty)$ so that for every box $B \subset \mathbb{R}^3$

$$prob\Big(\mathbf{v} \in B\Big) = \int_B f(\mathbf{v})\, d\mathbf{v} \ .$$

   We have the normalization

$$\int_{\mathbb{R}^3} f(\mathbf{v})\, d\mathbf{v} = 1 \ . \tag{14.9}$$

2. For $j = 1, 2, 3$ there is a smooth function $g_j : \mathbb{R} \to (0, \infty)$ so that, for every interval $I \subset \mathbb{R}$:

$$prob\Big(v_j \in I\Big) = \int_I g_j(z)\, dz \ .$$

3. For reasons of symmetry:

$$g_1 = g_2 = g_3 =: g \ .$$

4. (Independence) If $B = I_1 \times I_2 \times I_3$ then

$$prob\Big(\mathbf{v} \in B\Big) = prob\Big(v_1 \in I_1\Big) prob\Big(v_2 \in I_2\Big) prob\Big(v_3 \in I_3\Big) \ .$$

5. There is a smooth function $\Phi : [0, \infty) \to (0, \infty)$ with

$$f(\mathbf{v}) = \Phi(|\mathbf{v}|^2) \quad \text{where} \quad |\mathbf{v}|^2 = v_1^2 + v_2^2 + v_3^2 =: v^2 \ .$$

3.2. Derivation of Maxwell's Distribution

To draw conclusions from these assumptions, fix a vector $\mathbf{w} = (w_1, w_2, w_3) \in \mathbb{R}^3$ and let $h > 0$. Let B_h denote the box

$$B_h = [w_1, w_1 + h] \times [w_2, w_2 + h] \times [w_3, w_3 + h] \ .$$

Then we have

$$\begin{aligned} prob\Big(\mathbf{v} \in B_h\Big) &= \int_{B_h} \Phi(|\mathbf{v}|^2)\, d\mathbf{v} \\ &= \int_{w_1}^{w_1+h} g(z)dz \int_{w_2}^{w_2+h} g(z)dz \int_{w_3}^{w_3+h} g(z)dz \end{aligned}$$

Divide by h^3 and let $h \to 0$ to obtain

$$\Phi(|\mathbf{w}|^2) = g(w_1)g(w_2)g(w_3) \quad \text{for all} \quad \mathbf{w} \in \mathbb{R}^3 \; . \tag{14.10}$$

By choosing vectors $\mathbf{w}$ of particular form, we now exploit the above equation. First, let $\mathbf{w} = (z, 0, 0)$ to obtain

$$\Phi(z^2) = g(z)g^2(0) \; .$$

This shows that

$$g(z) = \frac{\Phi(z^2)}{g^2(0)} \; . \tag{14.11}$$

Second, let $\mathbf{w} = (w_1, w_2, 0), y = w_1^2, z = w_2^2$ and obtain

$$\Phi(y + z) = g(w_1)g(w_2)g(0) = \frac{\Phi(y)}{g^2(0)} \cdot \frac{\Phi(z)}{g^2(0)} \cdot g(0) \; . \tag{14.12}$$

(To get the first equation, use (14.10); to get the second equation, use (14.11).) Equation (14.12) reads:

$$\Phi(y + z) = \frac{\Phi(y)\Phi(z)}{g^3(0)} \quad \text{for} \quad y, z \in \mathbb{R} \; . \tag{14.13}$$

This formula for the function Φ is a major step in the derivation of Maxwell's distribution. As we will prove, equation (14.13) implies that the function Φ is an exponential.

Consider formula (14.13) and differentiate in y:

$$\frac{d}{dy}\Phi(y + z) = \frac{\Phi'(y)\Phi(z)}{g^3(0)} \quad \text{for} \quad y, z \in \mathbb{R} \; .$$

Set $y = 0$:

$$\Phi'(z) = -K\Phi(z) \quad \text{with} \quad K = -\frac{\Phi'(0)}{g^3(0)} \; .$$

Therefore, $\Phi(z)$ is an exponential:

$$\Phi(z) = C^3 e^{-Kz} \; . \tag{14.14}$$

The normalization condition (14.9) determines C in terms of $K > 0$:

$$\begin{aligned} 1 &= \int_{\mathbb{R}^3} \Phi(|\mathbf{v}|^2)\, d\mathbf{v} \\ &= C^3 \int_{\mathbb{R}^3} e^{-K|\mathbf{v}|^2}\, d\mathbf{v} \\ &= C^3 \Big(\int_{-\infty}^{\infty} e^{-Kz^2}\, dz \Big)^3 \\ &= (2C)^3 \Big(\int_0^{\infty} e^{-Kz^2}\, dz \Big)^3 \end{aligned}$$

Since

$$\int_0^\infty e^{-Kz^2}\, dz = \frac{1}{2}\sqrt{\pi/K}$$

one finds that

$$C = \sqrt{K/\pi}\ . \tag{14.15}$$

To summarize, from (14.14) and (14.15) we conclude that Maxwell's distribution function has the form

$$f(\mathbf{v}) = \Phi(v^2) = \Big(\frac{K}{\pi}\Big)^{3/2} e^{-Kv^2} \quad \text{for} \quad \mathbf{v} \in \mathbb{R}^3, \quad v = |\mathbf{v}|\ , \tag{14.16}$$

where $K > 0$ is a free parameter.

3.3. Relation between K and T

We now use (14.16) to compute the expected value of v^2:

$$\begin{aligned} \langle v^2 \rangle &= \int_{\mathbb{R}^3} v^2 f(\mathbf{v})\, d\mathbf{v} \\ &= (K/\pi)^{3/2} \int_{\mathbb{R}^3} v^2 e^{-Kv^2}\, d\mathbf{v} \\ &= (K/\pi)^{3/2} 4\pi \int_0^\infty v^4 e^{-Kv^2}\, dv \quad (\text{set } Kv^2 = \xi^2) \\ &= (K/\pi)^{3/2} 4\pi K^{-5/2} \int_0^\infty \xi^4 e^{-\xi^2}\, d\xi \\ &= \frac{3}{2K} \end{aligned}$$

The equation $\langle v^2 \rangle = \frac{3}{2K}$ and the relation $\frac{m}{2}\langle v^2 \rangle = \frac{3}{2}kT$ yield

$$\frac{3}{2}kT = \frac{3m}{4K} ,$$

i.e.,

$$K = \frac{m}{2kT} .$$

Maxwell's formula (14.16) becomes

$$f(\mathbf{v}) = \Phi(v^2) = \Big(\frac{m}{2\pi kT}\Big)^{3/2} e^{-mv^2/2kT} \quad \text{for} \quad \mathbf{v} \in \mathbb{R}^3 . \tag{14.17}$$

It is common to use still another function, depending directly on the speed v,

$$F(v) = 4\pi v^2 f(\mathbf{v}) = \sqrt{\frac{2}{\pi}} \Big(\frac{m}{kT}\Big)^{3/2} v^2 e^{-mv^2/2kT} . \tag{14.18}$$

This can be motivated as follows: If $v^{(0)}$ and $v^{(1)}$ denote two speeds with

$$0 \le v^{(0)} < v^{(1)}$$

then the probability of a velocity vector $\mathbf{v} \in \mathbb{R}^3$ to satisfy

$$v^{(0)} \le |\mathbf{v}| \le v^{(1)} \tag{14.19}$$

is

$$\begin{aligned} prob\Big(v^{(0)} \le |\mathbf{v}| \le v^{(1)}\Big) &= \int_A f(\mathbf{v})\, d\mathbf{v} \\ &= 4\pi \int_{v^{(0)}}^{v^{(1)}} v^2 \Phi(v^2)\, dv \\ &= \int_{v^{(0)}}^{v^{(1)}} F(v)\, dv \end{aligned}$$

Here the region $A \subset \mathbb{R}^3$ consists of all vectors $\mathbf{v}$ satisfying (14.19).

We summarize:

Theorem 14.1. *Let the function $F(v)$ be defined by (14.18) for $v \ge 0$. Let T denote the absolute temperature of a gas and let k denote Boltzmann's constant. Under Maxwell's assumptions, the velocity distribution satisfies:*

$$prob\Big(v^{(0)} \le |\mathbf{v}| \le v^{(1)}\Big) = \int_{v^{(0)}}^{v^{(1)}} F(v)\, dv .$$

The function $F(v)$ given in (14.18) is called the Maxwell–Boltzmann distribution.

Some implications: The most likely speed v_p is the speed at which the function $F(v)$ attains its maximum. One obtains

$$v_p = \sqrt{\frac{2kT}{m}} \; .$$

4. Optimal Heat Engines: Carnot's Cycle

Energy comes in different forms. The gas in our car contains chemical energy. In the motor, the chemical energy is first turned into heat energy, which is then turned into mechanical energy to move our car. (The heat is used for an expansion process in the motor, which is then translated into mechanical work.)

While it is easy to turn mechanical energy into heat through friction, it is less trivial to go the opposite direction. First, one must realize that one needs a *temperature difference* if one wants to turn heat energy into mechanical work. Otherwise, we could use the heat energy of the ocean water to move ships. Roughly, when heat flows from hot to cold, we can use a part of the high temperature heat energy to produce mechanical work. Any device which does this is called a heat engine.

In Section 4.1. we recall some basics about heat and mechanical energy. In Section 4.2. we introduce the instructive (V, p)–diagrams and explain the fundamental relation $dW = p\,dV$, which says that an increase of volume by dV at pressure p produces the mechanical work $dW = p\,dV$. We then use this to compute the work done by a heat engine during an isothermal expansion. We also derive the relation $pV^\gamma = const$ during an adiabatic expansion. These are preparations to understand the details of the Carnot cycle, which we describe in Section 4.3..

Throughout, for simplicity, we assume that we have an ideal gas in our heat engine. As we will explain at the end of Section 4.3., this restriction is not really important, but it makes the computations easier.

4.1. Heat Energy and Mechanical Energy

The most familiar forms of energy are heat energy and mechanical energy. The unit of heat energy is one *calorie*, which is the heat energy which increases the temperature of one gram of water by one degree Celsius. The unit of mechanical energy is

$$1\,Joule = 1\,Newton\,meter = 1\,\frac{kg\,m^2}{sec^2} \; .$$

Here $1Newton = 1kg\,m/sec^2$ is the unit of force, which is defined as the force that accelerates a mass of one kilogram by one meter per second^2. More concretely, the force $1Newton$

is about the weight of an apple. Thus, we can think of $1 Joule$ as the mechanical energy which lifts an apple by one meter against gravity.

It is easy to turn mechanical energy into heat through friction and, by careful experiments, the English physicist James Prescott Joule (1818–1889) found that 1 *cal* heat energy corresponds to 4.185 *Joule* mechanical energy.

4.2. Carnot's Heat Engine: Preparations

How can one turn heat energy into mechanical work? Sadi Carnot described an idealized heat engine which does this job and which is very simple. Remarkably, Carnot's heat engine has optimal efficiency, as we will explain in Section 4.3..

Let's first give a rough outline of what Carnot's engine does. The details will be given in the next section. The engine consists of a container of volume $V = V(t)$ that can change. Inside the container is a gas. When the engine is in contact with a hot reservoir of temperature T_1, the gas absorbs the heat energy Q_1. It then produces mechanical work E_1 and releases heat energy Q_2 towards the surroundings (or another reservoir) at the lower temperature T_2. By energy conservation,

$$Q_1 = E_1 + Q_2 \; .$$

The efficiency of the engine is, by definition,

$$\eta = \frac{E_1}{Q_1} \; .$$

The pressure of the gas in the container is denoted by $p = p(t)$. The pair $(V, p) = (V(t), p(t))$, then, describes the state of the machine at any given time t.

For simplicity, we will assume that we have an ideal gas in the container, which means that the absolute temperature T and the internal energy U of the gas obey the laws [4]

$$pV = RT, \quad U = CT \tag{14.20}$$

with positive constants R and C. Here the constants R and C will depend on the gas, but will be independent of the state (V, p) of the gas. (In a bouncing ball model, one can think of the internal energy U as the total kinetic energy of the moving and vibrating balls.)

As the state $(V(t), p(t))$ changes in time, the point $(V(t), p(t))$ moves in a (V, p)–diagram. For example, assume that at time t_1 the state is

$$(V_1, p_1) = (V(t_1), p(t_1))$$

and that in the time interval $t_1 \leq t \leq t_2$ the volume expands at constant temperature T_1 to

[4] The constant denoted by nR in (14.6) is simply denoted by R here.

$$(V_2, p_2) = (V(t_2), p(t_2)) \ .$$

By the ideal gas law $pV = RT$ we have

$$p(t) = \frac{RT_1}{V(t)} \ ,$$

which implies that the point $(V(t), p(t))$ moves along a hyperbola in the (V, p)–diagram. See Figure 14.2.

Infinitesimal Expansion and Mechanical Work. If a gas expands, it produces mechanical work. Precisely, if the volume increases by dV at pressure p, then the work performed is[5]

$$dW = p\,dV \ . \tag{14.21}$$

To explain this formula, we refer to Figure 14.1.

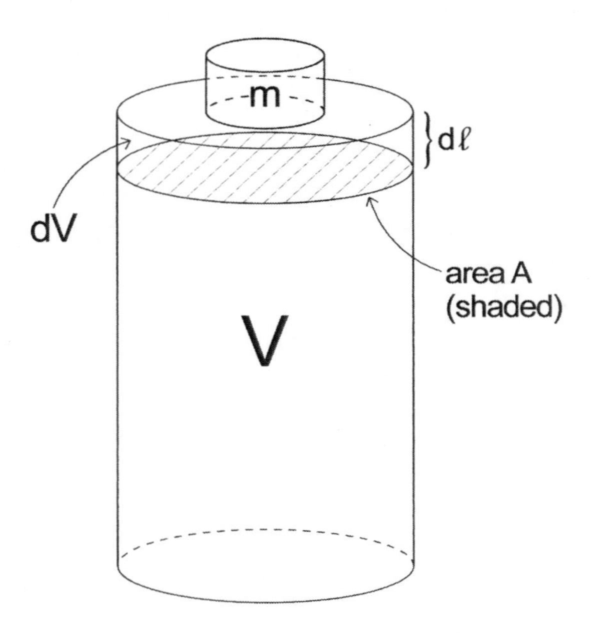

Figure 14.1. The volume V increases by dV

[5]One considers here an *infinitesimal* change of volume because one wants the pressure to remain unchanged during the expansion. To be precise, the pressure only changes infinitesimally when the volume changes from V to $V + dV$ and this infinitesimal change of p does not alter the outcome $dW = p\,dV$. The term $dp\,dV$ is negligible.

Suppose at time t the gas occupies the volume V at pressure p. Let A denote the area of the top of the container. The pressure p balances the weight mg, thus (recall that pressure is force per area):

$$p = \frac{mg}{A} .$$

Now assume that the weight mg is lifted by dl, leading to an increase of volume by $dV = A\,dl$. The work performed is

$$dW = mg\,dl = pA\,dl = p\,dV ,$$

which illustrates the fundamental formula (14.21).

Isothermal Expansion. Assume that the state (V_1, p_1) changes to (V_2, p_2) with $V_2 > V_1$. If the temperature $T = T_1 = p_1V_1/R$ remains unchanged during the process, we have an isothermal expansion. As the volume V increases from V_1 to V_2, the pressure changes as a function of V,

$$p = p(V) = \frac{RT_1}{V}, \quad V_1 \le V \le V_2 .$$

To obtain the total work performed during the expansion, we must add up the infinitesimal changes $dW = p\,dV$. The total work then is

$$\begin{aligned} W &= \int_{V_1}^{V_2} p\,dV \\ &= \int_{V_1}^{V_2} \frac{RT_1}{V}\,dV \\ &= RT_1 \ln(V_2/V_1) \end{aligned} \tag{14.22}$$

Where does the energy to perform the work W come from? Since the temperature $T = T_1$ remains unchanged during the isothermal expansion, the internal energy $U = CT$ also remains unchanged and cannot contribute to the work W. The assumption that $T = T_1$ remains constant implies that the energy W equals the heat energy absorbed by the heat engine from its surroundings. We will come back to this point in the next section.

Adiabatic Expansion. Consider again a state change from (V_1, p_1) to (V_2, p_2) with $V_2 > V_1$, but now assume that the heat engine does not absorb any heat energy (or other energy) from its surroundings. The state change is called *adiabatic*.

The machine performs mechanical work since the volume increases from V_1 to V_2, and for an adiabatic expansion the energy for this work will reduce the internal energy U, leading to a temperature drop of the engine. During the adiabatic process, the variables

$$V = V(t), \quad p = p(t), \quad T = T(t), \quad U = U(t)$$

change as a function of time and by (14.20) we have

$$p(t)V(t) = RT(t) = \frac{R}{C}U(t) \ .$$

Differentiation yields

$$p'V + pV' = \frac{R}{C}U' \ . \tag{14.23}$$

Because of (14.21) we have

$$dU = -p\,dV \ ,$$

thus

$$U' = -pV' \ .$$

Equation (14.23) then yields

$$p'V + pV' = -\frac{R}{C}pV' \ ,$$

thus

$$p'V + \gamma pV' = 0 \quad \text{with} \quad \gamma = 1 + \frac{R}{C} \ . \tag{14.24}$$

From this equation we find that

$$\frac{p'}{p} + \gamma\frac{V'}{V} = 0$$

and, therefore,

$$\begin{aligned} \frac{d}{dt}(pV^\gamma) &= p'V^\gamma + p\gamma V^{\gamma-1}V' \\ &= pV^\gamma\Big(\frac{p'}{p} + \gamma\frac{V'}{V}\Big) \\ &= 0 \end{aligned}$$

Our result says that the expression pV^γ remains unchanged for an adiabatic expansion and, with small changes of our arguments, we see that same holds for an adiabatic contraction. Since $pV = RT$ we also have

$$const = pV^\gamma = pVV^{\gamma-1} = RTV^{\gamma-1} \ .$$

To summarize, we have shown that

$$TV^{\gamma-1} = const \quad \text{with} \quad \gamma = 1 + \frac{R}{C} \tag{14.25}$$

for any adiabatic process.

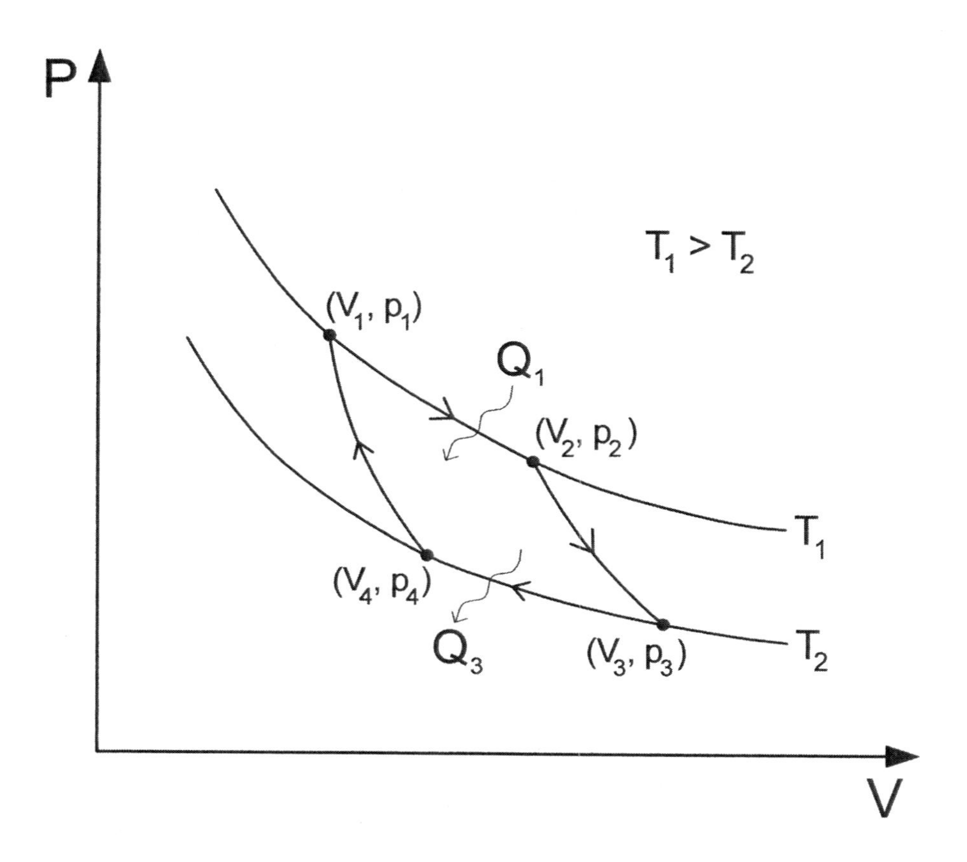

Figure 14.2. The Carnot cycle in a (V,p)–diagram

4.3. The Carnot Cycle and the Direction of Time

During the Carnot cycle, the heat engine goes through four processes. When the fourth process is completed, the machine is in its original state, but heat has been absorbed at the high temperature T_1, work has been performed, and heat has also been released at the low temperature T_2.

In each process, the state of the machine, specified by (V, p), changes slowly as a function of time. The point $(V(t), p(t))$ then moves along the curves shown in Figure 14.2, an illustration of the Carnot cycle due to Emile Clapeyron (1799–1864).

We will now go through the four processes and, at the end, compute the efficiency of Carnot's engine.

Process 1: Isothermal expansion from (V_1, p_1) to (V_2, p_2) at high temperature T_1.
According to formula (14.22), the work performed by the machine is

$$W_1 = RT_1 \ln(V_2/V_1) \quad \text{where} \quad p_1 V_1 = RT_1 \ .$$

As was explained after formula (14.22), the energy for the work W_1 equals the heat energy Q_1 absorbed by the machine at the temperature T_1 from the heat reservoir. In terms of energy,

$$W_1 = Q_1 \ .$$

Process 2: Adiabatic expansion from (V_2, p_2) to (V_3, p_3); temperature drop from T_1 to T_2.

There is no heat exchange of the machine with its surrounding during this process. As explained in the previous section, the temperature drops from T_1 to a lower temperature T_2 and, by formula (14.25), we have

$$T_2 V_3^{\gamma-1} = T_1 V_2^{\gamma-1} \ . \tag{14.26}$$

The mechanical work performed equals the drop in internal energy:

$$W_2 = U_2 = C(T_1 - T_2) \ .$$

Process 3: Isothermal compression from (V_3, p_3) to (V_4, p_4) at low temperature T_2.
For this process, mechanical work W_3 has to be provided, namely

$$W_3 = RT_2 \ln(V_3/V_4) \ .$$

The computation is the same as for Process 1. Since $V_3 > V_4$ we have $W_3 > 0$. This is the mechanical work which has to be provided to the machine from the outside.

Since the compression occurs at a constant temperature, the internal energy does not change during Process 3. Therefore, the energy W_3 equals the heat energy released by the machine to its surroundings at the low temperature T_2.

Process 4: Adiabatic compression from (V_4, p_4) to (V_1, p_1); temperature rise from T_2 to T_1.

There is no heat exchange of the machine with its surrounding during this process. By formula (14.25) we have

$$T_2 V_4^{\gamma-1} = T_1 V_1^{\gamma-1} \ . \tag{14.27}$$

During this process, the internal energy rises by

$$U_4 = C(T_1 - T_2)$$

and this energy has to be provided to the machine from the outside in terms of mechanical energy.

Conclusions. This completes our description of the four processes of the Carnot cycle and we will now draw some conclusions. Let us compute the total output of mechanical work. In this regard, Process 4 cancels Process 2 since Process 2 provides the work $C(T_1 - T_2)$, but Process 4 takes it back. The work provided by Process 1 is

$$W_1 = RT_1 \ln(V_2/V_1)$$

and Process 3 needs the works

$$W_3 = RT_2 \ln(V_3/V_4) \ .$$

Dividing equation (14.26) by equation (14.27) we obtain that

$$V_3/V_4 = V_2/V_1 \ .$$

Therefore, the total work performed is

$$W = W_1 - W_3 = R(T_1 - T_2) \ln(V_1/V_2) \ .$$

Now recall from Process 1 that

$$W_1 = Q_1 = RT_1 \ln(V_2/V_1)$$

is the heat energy absorbed by the machine at the high temperature T_1. This yields the following simple formula for the efficiency of Carnot's machine:

$$\eta = \frac{W_1 - W_3}{W_1} = \frac{T_1 - T_2}{T_1} \ . \tag{14.28}$$

Here is the main point of Carnot's engine: When it is run very slowly, it only goes through states of thermodynamic equilibrium. Therefore, it can run forward, as we have described it, but also backward. When run forward, the machine performs the mechanical work

$$W = W_1 - W_3 = Q_1 - W_3 \ ,$$

absorbs the heat energy Q_1 at the high temperature T_1, and releases the energy

$$Q_3 = W_3$$

at the low temperature T_2.

A **summary** of Carnot's engine, run forward, with slightly different notation is this:

a) The engine absorbs the heat energy Q_{in} at the high temperature T_1.

b) It splits the energy Q_{in} into

$$Q_{in} = W + Q_{out}$$

where W is the mechanical work performed and Q_{out} is released as heat energy at the low temperature T_2.

The efficiency of the engine is, by definition,

$$\eta = \frac{W}{Q_{in}} \ .$$

Assume we run Carnot's engine backwards. Then we put in the mechanical work W, take the heat energy Q_{out} from the low temperature reservoir (at T_2) and release the total energy $Q_{in} = W + Q_{out}$ at the high temperature T_1.

Now suppose there would be a heat engine more efficient than Carnot's working between the temperatures T_1 and T_2 with $T_1 > T_2$. The more efficient engine takes in the heat energy Q_1 at the high temperature T_1, splits it into

$$Q_{in} = \tilde{W} + \tilde{Q}_{out}$$

where $\tilde{W}$ is the mechanical work performed and $\tilde{Q}_{out}$ is the heat energy released at the low temperature T_2. By assumption, the machine has higher efficiency than Carnot's engine, thus

$$\tilde{W} > W \quad \text{and} \quad \tilde{Q}_{out} < Q_{out} \ .$$

If we now combine this hypothetical machine with Carnot's engine run backwards, we obtain the mechanical work

$$\tilde{W} - W > 0$$

by using the heat energy

$$Q_{out} - \tilde{Q}_{out} > 0$$

from the low temperature reservoir (or the low temperature environment). Thus, with this combination, we could take heat energy from the cold and get mechanical work out of it!

That this is practically impossible is intuitively obvious, as was already recognized by Carnot. The argument shows that Carnot's engine has optimal efficiency. In fact, with the same arguments, *any* heat engine which goes only through thermodynamic equilibria and thus can run forwards and backwards, has the same efficiency as Carnot's, namely

$$\eta = \frac{T_1 - T_2}{T_2} \ .$$

This explains our earlier remark about an ideal gas in Carnot's engine. It is not important that the gas obeys the ideal gas law (14.20). Real gases satisfy more complicated equations of state, but the resulting efficiency is the same, $\eta = (T_1 - T_2)/T_1$

As we have mentioned, it is intuitively obvious that one cannot take heat energy from the cold and use it to produce mechanical work. This insight can be formalized as the second law of thermodynamics. In 1850, it led Rudolf Clausius (1822–1888) to introduce the concept of entropy. Using entropy, one can quantify the observed irreversibility of natural phenomena.

We should mention here that in our description of the Carnot cycle we did not follow the historical path. Neither conservation of energy nor absolute temperature were understand in Carnot's time. The absolute temperature scale was introduced by William Thomson (Lord Kelvin, 1824–1907) around 1848. In fact, Kelvin realized that Carnot's ideas could be used to *define* a temperature scale, which he called absolute temperature, that did not depend on the properties of any particular substance. Absolute temperature is based on more fundamental physics than the Celsius scale, which depends on the properties of water.

It is an interesting subject to treat the second law of thermodynamics and the observed irreversibility of nature using bouncing molecules as a model. This is the starting point of statistical thermodynamics, whose originator was Ludwig Boltzmann (1844–1906). The relation between entropy S and probability W can be expressed by Boltzmann's formula

$$S = k \cdot logW \ ,$$

where k is Boltzmann's constant. An increase of entropy corresponds to an increase of disorder.

Illustration. To illustrate the formula

$$\eta = \frac{T_1 - T_2}{T_1}$$

for the efficiency of Carnot's engine, assume that

$$T_1 = 373\,K, \quad T_2 = 273\,K \ , \tag{14.29}$$

i.e., T_1 is the boiling temperature and T_2 is the freezing temperature of water. The efficiency is

$$\eta = \frac{100}{373} = 0.2681 \ .$$

This means that any heat engine working between the temperatures (14.29) has at best an efficiency of about a quarter. Only maximally (about) a quarter of the heat energy provided at the high temperature $T_1 = 373\,K$ can be turned into mechanical work.

In applications, the lower temperature T_2 is often the temperature of the surroundings. Then, to get high efficiencies, one has to increase the temperature T_1. This is a motivation for the diesel engine.

Carnot's engine is an *idealized* engine and practical machines do not reach the efficiency $(T_1 - T_2)/T_2$. In fact, to obtain 30% of Carnot efficiency is already a nontrivial engineering problem.

Chapter 15

Random Evolution: From Order to Chaos

Summary: What we observe in nature is typically irreversible although the *fundamental* laws of physics are believed to be time reversible. The observed irreversibility, then, may be a consequence of the very special initial condition in which the universe started.

In this chapter we illustrate this idea by a discrete–time random evolution process. Initially, the system is in an ordered state, but becomes disordered as time progresses. The way back to the ordered state is extremely unlikely, which expresses irreversibility.

The idea for the process goes back to the Austrian physicist and mathematician Paul Ehrenfest (1880–1933), who was a student of Ludwig Boltzmann (1844–1906). The model that we consider is slightly more difficult than Ehrenfest's original urn model.

1. Random Evolution of Macro States

Let N denote a positive integer. The value $N = 100$ is a good choice for numerical simulations, but you may also think of $N = 10^{23}$, close to Avogadro's constant.

Assume that you have N white balls $W_1, W_2, \ldots, W_N$ and N black balls $B_1, B_2, \ldots, B_N$. You also have two containers, a left one L and a right one R. (In this context, the containers are often called urns, which goes back to Ehrenfest.) At time $t = 0$ all white balls are in L and all black balls are in R. This is the ordered state at time zero.

At random, we draw a ball in L and a ball in R, giving each ball the same probability. We then exchange the two balls. It is clear, then, that at time $t = 1$ we have $N - 1$ white balls and one black ball in L; also, there are $N - 1$ black balls and one white ball in R. We then continue the process: At each time step, we randomly draw a ball in L and a ball in R and exchange them.

An example of L and R at time $t = 1$ for $N = 12$ is shown in Figure 15.1.

Our intuition tells us that, in the long run, there will be close to $N/2$ white balls and

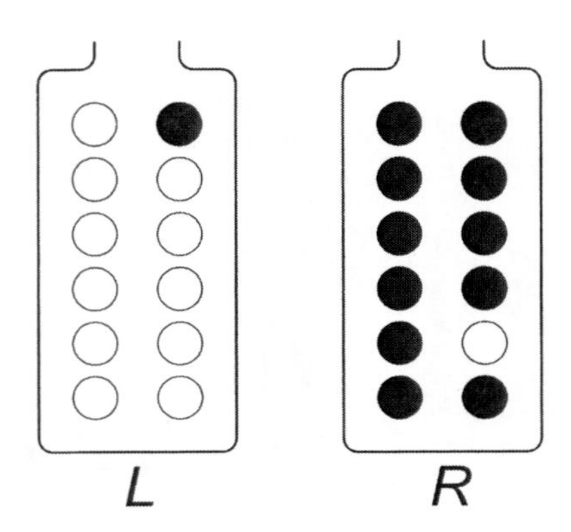

Figure 15.1. The urns L and R at time $t = 1$

$N/2$ black balls in both containers, in L as well as in R. Why is this so? Why is this disordered distribution much more likely than the ordered state at time zero? Why does random evolution increase the disorder?

The aim of this chapter is to explain and quantify this. It is possible, but — for large N — quite unlikely to ever get back to the ordered state unless one runs the random evolution for a very, very long time. For how long? And how does this time depend on N?

Before we can answer these questions, we will formalize the random evolution process, which is a Markov chain; see Section 4..

Formal Description: Let us denote by

$$j = X_t \in \{0, 1, \ldots, N\}$$

the number of white balls in the left container L at time t. The number $j = X_t$ is our random variable moving in the state space $\{0, 1, \ldots, N\}$. We think of the number $j = X_t$ as a description of the *macro state* of the system. If we know the number $j = X_t$ then we know that there are j white balls and $N - j$ black balls in L. Correspondingly, there are $N - j$ white balls and j black balls in R.

We do *not* know, however, *which* of the N white balls are in L. For this reason, the number $j = X_t$ describes the *macro state* of the system. In the next section, we will look at the *micro state*, which specifies precisely *which* of the j white balls and which of the $N - j$ black balls are in L.

Let us calculate the probabilities for the random evolution step

$$j = X_t \to i = X_{t+1} \ . \tag{15.1}$$

At time t there are j white balls and $N-j$ black balls in L, and we draw one ball randomly, giving each ball equal probability. With probability

$$p_j = \frac{j}{N} \tag{15.2}$$

we will draw a white ball in L and with probability

$$q_j = \frac{N-j}{N} = 1 - p_j \tag{15.3}$$

we will draw a black ball in L. Similarly, we randomly draw a ball in R, and it will be white with probability q_j and black with probability p_j.

Here is a list of the four possible drawings, their probabilities, and the resulting value for X_{t+1}:

L	R	probability	$i = X_{t+1}$
W	W	$p_j q_j$	j
B	W	q_j^2	$j+1$
W	B	p_j^2	$j-1$
B	B	$q_j p_j$	j

For example, with probability q_j we draw a black ball in L; also, independently, with probability q_j we draw a white ball in R. If both these drawing occur, then — and only then — the number of white balls in L will increase by one as time progresses from t to $t+1$. This case corresponds to the second row in the table.

If $X_t = j$ then the value of X_{t+1} can only be j (with probability $2p_jq_j$) or $j+1$ (with probability q_j^2) or $j-1$ (with probability p_j^2). Therefore, the transition matrix

$$P = (p_{ij})_{0 \le i,j \le N}$$

which encodes the probabilities of the step $X_t \to X_{t+1}$, is tridiagonal:

$$P = \begin{pmatrix} 0 & p_1^2 & & & & \\ q_0^2 & 2p_1q_1 & p_2^2 & & & \\ & q_1^2 & 2p_2q_2 & \ddots & & \\ & & q_2^2 & \ddots & p_{N-1}^2 & \\ & & & \ddots & 2p_{N-1}q_{N-1} & p_N^2 \\ & & & & q_{N-1}^2 & 0 \end{pmatrix} \tag{15.4}$$

The matrix P is column stochastic since

$$p_j^2 + 2p_jq_j + q_j^2 = (p_j + q_j)^2 = 1 \ .$$

Figure 15.2 shows a realization of the random evolution of the number of white balls in L, starting from the ordered state where $X_0 = N = 100$.

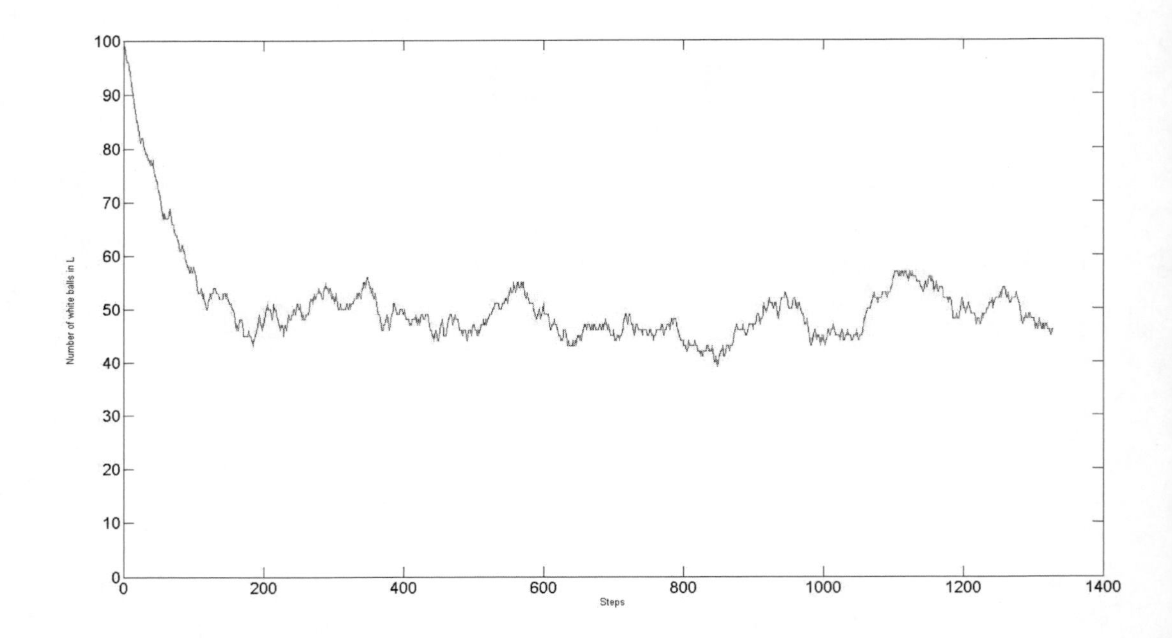

Figure 15.2. Realization of Random Evolution

We see that it takes somewhat less than $2N = 200$ time steps until X_t is about $N/2 = 50$. After time $t = 200$, the value of X_t fluctuates about the value $N/2 = 50$, but does not come close to $N = 100$ again. As time progressed, the ordered state $X_0 = 100$ has become disordered.

We will show in the next section that there exists a unique vector

$$\pi = \begin{pmatrix} \pi_0 \\ \pi_1 \\ \vdots \\ \pi_N \end{pmatrix} \in \mathbb{R}^{N+1}$$

satisfying

$$P\pi = \pi, \quad \sum_j \pi_j = 1, \quad \pi_j > 0 \quad \text{for} \quad j = 0, 1, \ldots, N \; . \tag{15.5}$$

This vector π gives us the stationary probability distribution of the random variable X_t. We will explain how one can obtain π: We will count how many micro states α_j lead to the macro state $j = X_t$, where j is the number of white balls in L. The probability π_j then is proportional to α_j.

The generalization of this counting process turns out to be an important principle of statistical physics: All micro states at the same energy level are equally likely. The probability of a particular macro state is proportional to the number of micro states leading to the macro state.

2. Macro States and Micro States

In our model, the N white balls and the N black balls are numbered:

$$W_1, \ldots, W_N \quad \text{and} \quad B_1, \ldots, B_N \ .$$

As in the previous section, let us assume that at time t there are $j = X_t$ white balls and $N - j$ black balls in L. The number j characterizes the macro state of the system. If we want to know its micro state, we must know exactly which of the j white balls and which of the $N - j$ black balls lie in L. Then, with this notion of a micro state, how many micro states correspond to the macro state $j = X_t$?

We will use the following result of combinatorics:

Lemma 15.1. *Let $0 \leq j \leq N$ and let $\mathcal{M}$ denote a set with N elements. The set $\mathcal{M}$ has* [1]

$$\binom{N}{j} = \frac{N!}{j!(N-j)!} = \frac{N(N-1)\cdots(N+1-j)}{j!}$$

subsets with j elements.

Proof: Let

$$\mathcal{M} = \{m_1, m_2, \ldots, m_N\} \ .$$

If we want to obtain a subset

$$\mathcal{S} = \{s_1, s_2, \ldots s_j\}$$

of $\mathcal{M}$ with j elements, then we have N choices for s_1, we have $N - 1$ choices for s_2, etc. This gives us

$$N(N-1)\cdots(N+1-j)$$

choices for the ordered j–tuple

[1] Read N choose j for the binomial coefficient $\binom{N}{j}$.

$$(s_1, s_2, \ldots, s_j)\ .$$

If $j = 2$ then the two ordered pairs

$$(s_1, s_2) \quad \text{and} \quad (s_2, s_1)$$

lead to the same set $\mathcal{S}$. Jf $j = 3$ then the $6 = 3!$ ordered 3–tuples

$$\begin{gathered}(s_1, s_2, s_3)\\(s_1, s_3, s_2)\\(s_2, s_1, s_3)\\(s_2, s_3, s_1)\\(s_3, s_1, s_2)\\(s_3, s_2, s_1)\end{gathered}$$

lead to the same set $\mathcal{S}$. One proves by induction that there are $j!$ orderings of each j–tuple $(s_1, s_2, \ldots, s_j)$. Therefore, the number of subsets $\mathcal{S}$ of $\mathcal{M}$ with j elements is

$$\frac{N(N-1)\cdots(N+1-j)}{j!} = \binom{N}{j}\ .$$

◇

Using this result, it is clear that there are $\binom{N}{j}$ choices of for the white balls and, independently, there are $\binom{N}{N-j}$ choices for the black balls leading to the macro state $j = X_t$. Therefore, the total number of micro states corresponding to the macro state $j = X_t$ is

$$\alpha_j := \binom{N}{j}\binom{N}{N-j} = \binom{N}{j}^2\ .$$

Let us denote by

$$M_N = \sum_{j=0}^{N} \alpha_j = \sum_{j=0}^{N} \binom{N}{j}^2$$

the total number of all micro states. It is reasonable to believe that — in the long run — all micro states will occur with equal probability. (In fact, this can be proved rigorously.) Then, since the proportion

$$\frac{\alpha_j}{M_N}$$

of micro states corresponds to the macro state $j = X_t$, we guess that — in the long run — the macro state $j = X_t$ occurs with probability

$$\pi_j = \frac{\alpha_j}{M_N} \quad \text{for} \quad j = 0, 1, \ldots, N \; . \tag{15.6}$$

We will prove that this guess is correct, i.e., the vector π with components (15.6) is the unique stationary probability density vector of the random process described in the previous section.

Theorem 15.1. *Recall the probabilities*

$$p_j = \frac{j}{N} \quad \text{and} \quad q_j = \frac{N-j}{N} \quad \text{for} \quad j = 0, 1, \ldots, N$$

and recall the transition matrix P given in (15.4). The vector $\pi \in \mathbb{R}^{N+1}$ with components (15.6) is the unique vector satisfying

$$P\pi = \pi, \quad \sum_j \pi_j = 1, \quad \pi_j > 0 \quad \text{for} \quad j = 0, 1, \ldots, N \; .$$

Proof: 1) We first prove that $P\pi = \pi$. For $1 \le j \le N-1$ the j–th row of P is

$$(0, \ldots, 0, q_{j-1}^2, 2p_jq_j, p_{j+1}^2, 0 \ldots, 0)$$

where the entry $2p_jq_j$ is on the diagonal of P. We must then show that

$$q_{j-1}^2\pi_{j-1} + 2p_jq_j\pi_j + p_{j+1}^2\pi_{j+1} = \pi_j \; . \tag{15.7}$$

Multiplying (15.7) by N^2M_N the equation becomes

$$(N+1-j)^2 \binom{N}{j-1}^2 + 2j(N-j)\binom{N}{j}^2 + (j+1)^2\binom{N}{j+1}^2 = N^2\binom{N}{j}^2 \; .$$

Now divide by $\binom{N}{j}^2$ to obtain the equivalent equation

$$(N+1-j)^2\frac{j^2}{(N+1-j)^2} + 2j(N-j) + (j+1)^2\frac{(N-j)^2}{(j+1)^2} = N^2 \; .$$

The last equation is true since

$$j^2 + 2j(N-j) + (N-j)^2 = (j + N - j)^2 = N^2 \; .$$

Thus we have shown that $(P\pi)_j = \pi_j$ for $1 \le j \le N$. The special cases $j = 0$ and $j = N$ are easy to check, and the vector equation $P\pi = \pi$ is proved.

2) It remains to prove the uniqueness statement. Assume, then, that there exists a vector $\beta \in \mathbb{R}^{N+1}$, $\beta \neq \pi$, with

$$P\beta = \beta, \quad \sum_j \beta_j = 1, \quad \beta_j > 0 \quad \text{for} \quad j = 0, 1, \ldots, N \; .$$

If we consider the matrix P^2 it is easy to see that all its elements on the diagonal and on the four lines next to the diagonal are strictly positive. Taking higher powers P^3, P^4, etc. we see that all elements of the matrix

$$A = P^{N-1} = (a_{ij})$$

are strictly positive, $a_{ij} > 0$. It is also clear that

$$A\pi = \pi, \quad A\beta = \beta \; . \tag{15.8}$$

Now consider the half–line

$$\phi(s) := \pi - s\beta, \quad 0 \le s < \infty \; .$$

We can choose a parameter value s so that

$$\phi_i(s) \ge 0 \quad \text{for} \quad i = 0, \ldots, N; \quad \phi_j(s) = 0 \quad \text{for some } j \; . \tag{15.9}$$

The vector $\phi(s)$ is not the zero vector since otherwise π would be a multiple of β. Now apply the matrix $A = P^{N-1}$ to $\phi(s)$. Using (15.8) we see that

$$A\phi(s) = \phi(s) \; .$$

However, since all entries a_{ij} are strictly positive, all components of the vector $A\phi(s)$ are strictly positive. This contradicts $\phi(s) = A\phi(s)$ and $\phi_j(s) = 0$. The contradiction proves the uniqueness of the stationary probability density π. $\diamond$

3. Back to the Ordered State

Assume now, for simplicity, that N is even and that at time $t = 0$ there are $N/2$ white balls and $N/2$ black balls in L, i.e.,

$$X_0 = \frac{N}{2} \; .$$

If we run the random evolution for very long, will we ever reach the ordered state

$$X_t = N\ ?$$

More precisely, let us ask: How large will we have to choose the time T so that, with probability $\frac{1}{2}$, the ordered state

$$X_t = N$$

will occur at some time t with

$$t \in \{1, 2, \ldots, T\}\ ?$$

We want to get a rough estimate for T. First, as we have shown, the stationary probability of the state $X_t = N$ is

$$\pi_N = \frac{\alpha_N}{M_N} = \frac{1}{M_N}$$

with

$$M_N = \sum_{j=0}^{N} \alpha_j = \sum_{j=0}^{N} \binom{N}{j}^2 .$$

We know that

$$\sum_{j=0}^{N} \binom{N}{j} = (1+1)^N = 2^N$$

and

$$\begin{aligned} 2^N &\leq \sum_{j=0}^{N} \binom{N}{j}^2 \\ &\leq \Big(\sum_{j=0}^{N} \binom{N}{j}\Big)^2 \\ &\leq 2^{2N} \end{aligned}$$

thus

$$2^{-2N} \leq \pi_N \leq 2^{-N} . \tag{15.10}$$

If the time T is very large and we choose a time $t \in \{1, 2, \ldots, T\}$ at random, we expect that the state

$$X_t = N$$

will occur with probability π_N. Thus, X_t will be different from N with probability

$$1 - \pi_N \ .$$

If we assume (somewhat incorrectly) that the *events*

$$X_t = N$$

are all independent for $t = 1, 2, \ldots, T$, then the probability that X_t is different from N for all $t = 1, 2, \ldots, T$ equals

$$(1 - \pi_N)^T \ .$$

This leads us to choose T so that

$$(1 - \pi_N)^T = \frac{1}{2} \ .$$

Taking logarithms we find that

$$T \ln(1 - \pi_N) = -\ln 2 \ .$$

Since π_N is a very small positive number, we may use the approximation

$$\ln(1 - \pi_N) \sim -\pi_N \ ,$$

which yields

$$T \sim \frac{\ln 2}{\pi_N} \ .$$

From (15.10) we have

$$\frac{1}{\pi_N} \geq 2^N$$

and find that (approximately)

$$T \geq 2^N \ln 2 \ .$$

For $N = 100$ one obtains the lower bound

$$T_0 = 2^{100} \ln 2 \sim 10^{30} \ .$$

A computer which executes 10^{12} time steps $X_t \to X_{t+1}$ per second needs 10^{18} seconds to carry out 10^{30} time steps.

The time 10^{18} seconds is about 30 billion years, which is longer than the (estimated) age of the universe.

References

[1] L. Ya. Adrianova, *Introduction to Linezr Systems of Differential Equations*, Translations of Mathematical Monographs, Vol. 146, AMS, 1991.

[2] Friedrich Wilhem Bessel, *Untersuchung des Theils der planetarischen Störung, welcher aus der Bewegung der Sonne entsteht*, Berliner Abh., (1824), pp. 1-52.

[3] Earl Coddington and Norman Levinson, *Theory of Ordinary Differential Equations*, McGraw–Hill, 1955.

[4] Peter Colwell, *Solving Kepler's Equation over Three Centuries*, Willmann–Bell, 1993.

[5] Florin Diacu and Philip Holmes, *Celestial Encounters*, Princeton University Press, 1996.

[6] David J. Griffiths, *Introduction to Elementary Particles*, Wiley, John & Sons, 1987.

[7] Jack K. Hale and Hüseyin Kocak, *Dynamics and Bifurcations*, Springer–Verlag, 1991.

[8] Stephen Hawking and Leonard Mlodinow, *The Grand Design*, Bantam Books, 2010.

[9] Andrey Kolmogorov, *Grundbegriffe der Wahrscheinlichkeitsrechnung*, Springer–Verlag, 1933.

[10] Heinz–Otto Kreiss, *Problems with Different Time Scales*, Acta Numerica, Vol. 1, (1992), pp. 101–139.

[11] Heinz–Otto Kreiss and Jens Lorenz, *Resolvent Estimates and Quantification of Nonlinear Stability*, Acta Mathematica Sinica, English Series, Vol. 16, No. 1 (2000), pp. 1-20.

[12] Yuri Kuznetsov, *Elements of Applied Bifurcation Theory*, Applied Mathematical Sciences 112, Springer–Verlag, 1995.

[13] Makoto Matsumoto and Takuji Nishimura, *Mersenne twister: a 623–dimensionally equidistributed uniform pseudo–random number generator*, ACM Transactions on Modeling and Computer Simulation 8 (1), pp. 3–30.

[14] Richard Montgomery, *A New Solution to the Three–Body Problem*, Notices of the AMS, 48, No. 5, (2001), pp. 471-481.

[15] Erwin Schrödinger, *What is Life?*, Cambridge University Press, 1944.

[16] Steven H. Strogatz, *Nonlinear Dynamics and Chaos*, Westview Press, 2000.

[17] Sergio B. Volchan, *What is a Random Sequence?* The American Mathematical Monthly 109, (2002), pp. 46-63.

[18] Hermann Weyl, *Über die Gleichverteilung von Zahlen mod. Eins*, Math. Ann. 77, No. 3, (1916), pp. 313-352.

Index